A
Mathematician's
Lament

A Mathematician's Lament

PAUL LOCKHART

With a Foreword by Keith Devlin

BELLEVUE LITERARY PRESS
NEW YORK

First published in the United States in 2009 by
Bellevue Literary Press, New York

FOR INFORMATION, CONTACT:
Bellevue Literary Press
90 Broad Street
Suite 2100
New York, NY 10004
www.blpress.org

Library of Congress Cataloging-in-Publication Data
Lockhart, Paul.
    A mathematician's lament / Paul Lockhart; with a foreword by
Keith Devlin. – 1st ed.
    p. cm.
    Includes bibliographical references and index.
Contents: Lamentation. Mathematics and culture; Mathematics in
school; The mathematics curriculum; High school geometry:
instrument of the devil—Exultation.
    ISBN 978-1-93413-717-8 (pbk.)
    1. Mathematics—Study and teaching—United States—Evaluation.
2. Educational evaluation—United States.
QA13.L63 2009
510.71   2009011845

Bellevue Literary Press would like to thank all its generous
donors—individuals and foundations—for their support.

Book design and type formatting by Bernard Schleifer

Bellevue Literary Press is committed to ecological stewardship in our
book production practices, working to reduce our impact
on the natural environment.

♾ This book is printed on acid-free paper

Manufactured in the United States of America

FIRST EDITION

13   15   17   18   16   14   12

paperback ISBN 978-1-934137-17-8
ebook ISBN 978-1-934137-33-8

For Stanley, who asked me to write it.

If you want to build a ship, don't drum up people to collect wood and don't assign them tasks and work, but rather teach them to long for the endless immensity of the sea.

—Antoine de Saint Exupéry

Foreword

$\mathfrak{I}$N LATE 2007, AN AUDIENCE MEMBER AT A TALK I gave handed me a 25-page typewritten document called *A Mathematician's Lament*, saying he thought I might like it. Written by a mathematics teacher called Paul Lockhart, the essay had been circling somewhat erratically through the mathematics education community since its author first wrote it in 2002, but it had never been published. The audience member's prediction turned out to be an understatement. I loved it, and felt that the words of this Paul Lockhart—whoever he was—deserved a much wider audience. And so I did something I have never done before, and probably never will again: after tracking down the essay's author—not entirely straightforward since the essay bore no contact information—and securing his permission, I devoted an entire issue of my monthly online column "Devlin's Angle" on the Mathematical Association of America's web-zine *MAA Online* (www.maa.org) to reproducing the entire essay in its original form. It was the quickest and most effective way I knew to get it in front of the mathematics and mathematics education communities.

When *A Mathematician's Lament* appeared in my March 2008 column, I introduced it with these words:

> *It is, quite frankly, one of the best critiques of current K–12 mathematics education I have ever seen.*

I was expecting a strong response. What ensued was a firestorm. Paul's words struck a very, very loud chord that resonated around the world. In addition to many emails expressing appreciation, requests flooded in—many to me, since by agreement I did not publish Paul's contact information—for reproduction and translation rights. (The volume you have in your hands arose in precisely this way.)

It wasn't that Paul was saying something that

countless mathematicians and math teachers have not said before. Nor were the points he raised new to those in the sometimes divided world of mathematics education who wrote to disagree with much if not all of what he wrote. What was different was the eloquence of his words and the obvious passion he injected into them. This was not just good writing; this was *great* writing, coming right from the heart.

Make no mistake about it, *A Mathematician's Lament*, and this greatly expanded book version, is an *opinion* piece. Paul has strong views on how mathematics should be taught, and he argues forcefully for his approach, and against much of the status quo in today's world of school mathematics education. What singles him out, besides his personal and captivating writing style, is that he brings to the thorny and much-debated issues of mathematics education a perspective that few others are able to draw upon. Paul is one of those very rare birds who began as an accomplished professional research mathematician, teaching students in universities, and then realized his true calling was in K-12 teaching, which is the career he has followed for many years now.

In my view, this book, like the original essay it came from, should be obligatory reading for anyone

Foreword

going into mathematics education, for every parent of a school-aged child, and for any school or government official with responsibilities toward mathematics teaching. You may not agree with everything Paul says. You may think his approach to teaching is not one that every teacher could successfully adopt. But you should read what he says and reflect on his words. *A Mathematician's Lament* is already a recognized landmark in the world of mathematics education that cannot and should not be ignored. I am not going to tell you how I think you should respond. As Paul himself would agree, that is for every individual reader to do. But I will tell you this. I would have *loved* to have had Paul Lockhart as my school mathematics teacher.

KEITH DEVLIN
Stanford University

A
Mathematician's
Lament

PART I

Lamentation

𝕬 MUSICIAN WAKES FROM A TERRIBLE NIGHTMARE. In his dream he finds himself in a society where music education has been made mandatory. "We are helping our students become more competitive in an increasingly sound-filled world." Educators, school systems, and the state are put in charge of this vital project. Studies are commissioned, committees are formed, and decisions are made—all without the advice or participation of a single working musician or composer.

Since musicians are known to set down their ideas in the form of sheet music, these curious black dots and lines must constitute the "language of music." It is imperative that students become fluent in this language if they are to attain any degree of musical competence; indeed, it would be ludicrous to expect a child to sing a song or play an instrument without having a thorough grounding in music notation and theory. Playing and listening to music, let alone composing an original piece, are considered very advanced topics and are generally put off until college, and more often graduate school.

As for the primary and secondary schools, their mission is to train students to use this language—to jiggle symbols around according to a fixed set of rules: "Music class is where we take out our staff paper, our teacher puts some notes on the board, and we copy them or transpose them into a different key. We have to make sure to get the clefs and key signatures right, and our teacher is very picky about making sure we fill in our quarter-notes completely. One time we had a chromatic scale problem and I did it right, but the teacher gave me no credit because I had the stems pointing the wrong way."

In their wisdom, educators soon realize that even very young children can be given this kind of musical instruction. In fact it is considered quite shameful if one's third-grader hasn't completely memorized his circle of fifths. "I'll have to get my son a music tutor. He simply won't apply himself to his music homework. He says it's boring. He just sits there staring out the window, humming tunes to himself and making up silly songs."

In the higher grades the pressure is really on. After all, the students must be prepared for the standardized tests and college admissions exams. Students must take courses in scales and modes, meter, harmony, and counterpoint. "It's a lot for them to learn, but later in college when they finally get to hear all this stuff, they'll really appreciate all the work they did in high school." Of course, not many students actually go on to concentrate in music, so only a few will ever get to hear the sounds that the black dots represent. Nevertheless, it is important that every member of society be able to recognize a modulation or a fugal passage, regardless of the fact that they will never hear one. "To tell you the truth, most students just aren't very good at music. They are bored in class, their skills are terrible, and their homework is barely legible. Most of them couldn't care less about how important music is in today's world; they just want to take the minimum number of music courses and be done with it. I guess there are just music people and non-music people. I had this one kid, though, man was she sensational! Her sheets were impeccable—every note in the right place, perfect calligraphy, sharps, flats, just beautiful. She's going to make one hell of a musician someday."

Waking up in a cold sweat, the musician realizes, gratefully, that it was all just a crazy dream. "Of course," he reassures himself, "no society would ever reduce such a beautiful and meaningful art form to something so mindless and trivial; no culture could be so cruel to its children as to deprive them of such a natural, satisfying means of human expression. How absurd!"

Meanwhile, on the other side of town, a painter has just awakened from a similar nightmare . . .

. . . I was surprised to find myself in a regular school classroom—no easels, no tubes of paint. "Oh we don't actually apply paint until high school," I was told by the students. "In seventh grade we mostly study colors and applicators." They showed me a worksheet. On one side were swatches of color with blank spaces next to them. They were told to write in the names. "I like painting," one of the students remarked. "They tell me what to do and I do it. It's easy!"

After class I spoke with the teacher. "So your students don't actually do any painting?" I asked. "Well, next year they take Pre-Paint-by-Numbers," the teacher replied. "That prepares them for the main Paint-by-Numbers sequence in high school. So they'll get to use what they've learned here and apply it to real-life painting situations—dipping the brush into paint, wiping it off, stuff like that. Of course we track our students by ability. The really excellent painters—the ones who know their colors and brushes backwards and forwards—they get to the actual painting a little sooner, and some of them even take the Advanced Placement classes for college credit. But mostly we're just trying to give these kids a good foundation in what painting is all about, so when they get out there in the real world and paint their kitchen they don't make a total mess of it."

"Um, these high school classes you mentioned . . ."

"You mean Paint-by-Numbers? We're seeing much higher enrollments lately. I think it's mostly coming from parents wanting to make sure their kid gets into a good college. Nothing looks better than Advanced Paint-by-Numbers on a high school transcript."

" Why do colleges care if you can fill in numbered regions with the corresponding color?"

"Oh, well, you know, it shows clear-headed logical thinking. And of course if a student is planning to major in one of the visual sciences, like fashion or interior decorating, then it's really a good idea to get your painting requirements out of the way in high school."

"I see. And when do students get to paint freely, on a blank canvas?"

"You sound like one of my professors! They were always going on about expressing yourself and your feelings and things like that—really way-out-there abstract stuff. I've got a degree in painting myself, but I've never really worked much with blank canvasses. I just use the Paint-by-Numbers kits supplied by the school board."

* * *

Sadly, our present system of mathematics education is precisely this kind of nightmare. In fact, if I had to design a mechanism for the express purpose of *destroying* a child's natural curiosity and love of pattern-making, I couldn't possibly do as good a job as is currently being done—I simply wouldn't have the imagination to come up with the kind of senseless, soul-crushing ideas that constitute contemporary mathematics education.

Everyone knows that something is wrong. The politicians say, "We need higher standards." The schools say, "We need more money and equipment." Educators say one thing, and teachers say another. They are all wrong. The only people who understand what is going on are the ones most often blamed and least often heard: the students. They say, "Math class is stupid and boring," and they are right.

Mathematics and Culture

T HE FIRST THING TO UNDERSTAND IS THAT MATHE-matics is an art. The difference between math and the other arts, such as music and painting, is that our culture does not recognize it as such. Everyone understands that poets, painters, and musicians create works of art, and are expressing themselves in word, image, and sound. In fact, our society is rather generous when it comes to creative expression; architects, chefs, and even television directors are considered to be working artists. So why not mathematicians?

Part of the problem is that nobody has the faintest idea what it is that mathematicians do. The common perception seems to be that mathematicians are somehow connected with science—perhaps they help the scientists with their formulas, or feed big numbers into computers for some reason or other. There is no question that if the world had to be divided into the "poetic dreamers" and the "rational thinkers" most people would place mathematicians in the latter category.

Nevertheless, the fact is that there is nothing as dreamy and poetic, nothing as radical, subversive, and psychedelic, as mathematics. It is every bit as mind-blowing as cosmology or physics (mathematicians *conceived* of black holes long before astronomers actually found any), and allows more freedom of expression than poetry, art, or music (which depend heavily on properties of the physical universe). Mathematics is the purest of the arts, as well as the most misunderstood.

So let me try to explain what mathematics is, and what mathematicians do. I can hardly do better than to begin with G. H. Hardy's excellent description:

> A mathematician, like a painter or poet, is a maker of patterns. If his patterns are more permanent than theirs, it is because they are made with *ideas*.

So mathematicians sit around making patterns of ideas. What sort of patterns? What sort of ideas? Ideas about the rhinoceros? No, those we leave to the biologists. Ideas about language and culture? No, not usually. These things are all far too complicated for most mathematicians' taste. If there is anything like a unifying aesthetic principle in mathematics, it is this: *simple is beautiful*. Mathematicians enjoy thinking about the simplest possible things, and the simplest possible things are *imaginary*.

For example, if I'm in the mood to think about shapes—and I often am—I might imagine a triangle inside a rectangular box:

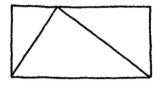

I wonder how much of the box the triangle takes up—two-thirds maybe? The important thing to understand is that I'm not talking about this *drawing* of a triangle in a box. Nor am I talking about some metal triangle forming part of a girder system for a bridge. There's no ulterior practical purpose here. I'm just *playing*. That's what math is—wondering, playing, amusing yourself with your imagination. For one thing, the question of how much of the box the triangle takes up doesn't even make any *sense* for real, physical objects. Even the most carefully made physical triangle is still a hopelessly complicated collection of jiggling atoms; it changes its size from one minute to the next. That is, unless you want to talk about some sort of *approximate* measurements. Well, that's where the aesthetic comes in. That's just not simple, and consequently it is an ugly question that depends on all sorts of real-world details. Let's leave that to the scientists. The *mathematical* question is about an imaginary triangle inside an imaginary box. The edges are perfect because I want them to be— that is the sort of object I prefer to think about. This is a major theme in mathematics: things are what you want them to be. You have endless choices; there is no reality to get in your way.

On the other hand, once you have made your choices (for example I might choose to make my triangle symmetrical, or not) then your new creations do what they do, whether you like it or not. This is the amazing thing about making imaginary patterns: they talk back! The triangle takes up a certain amount of its box, and I don't have any control over what that amount is. There is a number out there, maybe it's two-thirds, maybe it isn't, but I don't get to say what it is. I have to *find out* what it is.

So we get to play and imagine whatever we want and make patterns and ask questions about them. But how do we answer these questions? It's not at all like science. There's no experiment I can do with test tubes and equipment and whatnot that will tell me the truth about a figment of my imagination. The only way to get at the truth about our imaginations is to use our imaginations, and that is hard work.

In the case of the triangle in its box, I do see something simple and pretty:

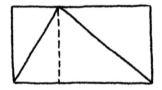

If I chop the rectangle into two pieces like this, I can see that each piece is cut diagonally in half by the sides of the triangle. So there is just as much space inside the triangle as outside. That means that the triangle must take up exactly half the box!

This is what a piece of mathematics looks and feels like. That little narrative is an example of the mathematician's art: asking simple and elegant questions about our imaginary creations, and crafting satisfying and beautiful explanations. There is really nothing else quite like this realm of pure idea; it's fascinating, it's fun, and it's free!

Now where did this idea of mine come from? How did I know to draw that line? How does a painter know where to put his brush? Inspiration, experience, trial and error, dumb luck. That's the art of it, creating these beautiful little poems of thought, these sonnets of pure reason. There is something so wonderfully transformational about this art form. The relationship between the triangle and the rectangle was a mystery, and then that one little line made it obvious. I couldn't see, and then all of a sudden I could. Somehow, I was able to create a profound simple beauty out of nothing, and change myself in the process. Isn't that what art is all about?

This is why it is so heartbreaking to see what is being done to mathematics in school. This rich and fascinating adventure of the imagination has been reduced to a sterile set of facts to be memorized and procedures to be followed. In place of a simple and natural question about shapes, and a creative and rewarding process of invention and discovery, students are treated to this:

Triangle Area Formula:
$A = ½ b h$

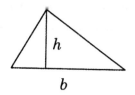

"The area of a triangle is equal to one-half its base times its height." Students are asked to memorize this formula and then "apply" it over and over in the "exercises." Gone is the thrill, the joy, even the pain and frustration of the creative act. There is not even a *problem* anymore. The question has been asked and answered at the same time—there is nothing left for the student to do.

Now let me be clear about what I'm objecting to. It's not about formulas, or memorizing interesting facts. That's fine in context, and has its place just as learning a vocabulary does—it helps you to create richer, more nuanced works of art. But it's not the *fact* that triangles take up half their box that matters. What matters is the beautiful *idea* of chopping it with the line, and how that might inspire other beautiful ideas and lead to creative breakthroughs in other problems—something a mere statement of fact can never give you.

By removing the creative process and leaving only the results of that process, you virtually guarantee that no one will have any real engagement with the subject. It is like saying that Michelangelo created a beautiful sculpture, without letting me see it. How am I supposed to be inspired by that? (And of course it's actually much worse than this—at least it's understood that there *is* an art of sculpture that I am being prevented from appreciating).

By concentrating on *what*, and leaving out *why*, mathematics is reduced to an empty shell. The art is not in the "truth" but in the explanation, the argument. It is the argument itself that gives the truth its context, and determines what is really being said and meant. Mathematics is *the art of explanation*. If you deny students the opportunity to engage in this activity—to pose their own problems, to make their own conjectures and discoveries, to be wrong, to be creatively frustrated, to have an inspiration, and to cobble together their own explanations and proofs— you deny them mathematics itself. So no, I'm not complaining about the presence of facts and formulas in our mathematics classes, I'm complaining about the lack of *mathematics* in our mathematics classes.

If your art teacher were to tell you that painting is all about filling in numbered regions, you would know

that something was wrong. The culture informs you
—there are museums and galleries, as well as the art
in your own home. Painting is well understood by
society as a medium of human expression. Likewise,
if your science teacher tried to convince you that
astronomy is about predicting a person's future based
on their date of birth, you would know she was
crazy—science has seeped into the culture to such an
extent that almost everyone knows about atoms and
galaxies and laws of nature. But if your math teacher
gives you the impression, either expressly or by
default, that mathematics is about formulas and definitions and memorizing algorithms, who will set you
straight?

The cultural problem is a self-perpetuating monster: students learn about math from their teachers,
and teachers learn about it from their teachers, so this
lack of understanding and appreciation for mathematics in our culture replicates itself indefinitely.
Worse, the perpetuation of this "pseudo-mathemat-
ics," this emphasis on the accurate yet mindless
manipulation of symbols, creates its own culture and
its own set of values. Those who have become adept
at it derive a great deal of self-esteem from their suc-
cess. The last thing they want to hear is that math is

really about raw creativity and aesthetic sensitivity. Many a graduate student has come to grief when they discover, after a decade of being told they were "good at math," that in fact they have no real mathematical talent and are just very good at following directions. Math is not about following directions, it's about making new directions.

And I haven't even mentioned the lack of mathematical criticism in school. At no time are students let in on the secret that mathematics, like any literature, is created by human beings for their own amusement; that works of mathematics are subject to critical appraisal; that one can have and develop mathematical *taste*. A piece of mathematics is like a poem, and we can ask if it satisfies our aesthetic criteria: Is this argument sound? Does it make sense? Is it simple and elegant? Does it get me closer to the heart of the matter? Of course there's no criticism going on in school—there's no art being done to criticize!

Why don't we want our children to learn to do mathematics? Is it that we don't trust them, that we think it's too hard? We seem to feel that they are capable of making arguments and coming to their own conclusions about Napoleon. Why not about triangles? I think it's simply that we as a culture don't know what mathematics is. The impression we are given is of something very cold and highly technical, that no one could possibly understand—a self-fulfilling prophesy if there ever was one.

It would be bad enough if the culture were merely ignorant of mathematics, but what is far worse is that people actually think they *do* know what math is about—and are apparently under the gross misconception that mathematics is somehow useful to society! This is already a huge difference between mathematics and the other arts. Mathematics is viewed by the culture as some sort of tool for science and technology. Everyone knows that poetry and music are for pure enjoyment and for uplifting and ennobling the human spirit (hence their virtual elimination from the public school curriculum), but no, math is *important*.

> SIMPLICIO: Are you really trying to claim that mathematics offers no useful or practical applications to society?

> SALVIATI: Of course not. I'm merely suggesting that just because something happens to have practical consequences doesn't mean that's what it is *about*. Music can lead armies into battle, but that's not why people write sym-

phonies. Michelangelo decorated a ceiling, but I'm sure he had loftier things on his mind.

SIMPLICIO: But don't we need people to learn those useful consequences of math? Don't we need accountants and carpenters and such?

SALVIATI: How many people actually use any of this "practical math" they supposedly learn in school? Do you think carpenters are out there using trigonometry? How many adults remember how to divide fractions, or solve a quadratic equation? Obviously the current practical training program isn't working, and for good reason: it is excruciatingly boring, and nobody ever uses it anyway. So why do people think it's so important? I don't see how it's doing society any good to have its members walking around with vague memories of algebraic formulas and geometric diagrams, and clear memories of hating them. It might do some good, though, to show them something beautiful and give them an opportunity to enjoy being creative, flexible, open-minded thinkers—the kind of thing a *real* mathematical education might provide.

SIMPLICIO: But people need to be able to balance their checkbooks, don't they?

SALVIATI: I'm sure most people use a calculator for everyday arithmetic. And why not? It's certainly easier and more reliable. But my point is not just that the current system is so terribly bad, it's that what it's missing is so wonderfully good! Mathematics should be taught as art for art's sake. These mundane "useful" aspects would follow naturally as a trivial by-product. Beethoven could easily write an advertising jingle, but his motivation for learning music was to create something beautiful.

SIMPLICIO: But not everyone is cut out to be an artist. What about the kids who aren't "math people"? How would they fit into your scheme?

SALVIATI: If everyone were exposed to mathematics in its natural state, with all the challenging fun and surprises that that entails, I think we would see a dramatic change both in the attitude of students toward mathematics, and in our conception of what it means to be good at math. We are losing so many potentially gifted mathematicians—creative, intelligent people who rightly reject what appears to be a meaningless and sterile subject. They are simply too smart to waste their time on such piffle.

SIMPLICIO: But don't you think that if math class were made more like art class that a lot of kids just wouldn't learn anything?

SALVIATI: They're not learning anything now! Better to not have math classes at all than to do what is currently being done. At least some people might have a chance to discover something beautiful on their own.

SIMPLICIO: So you would remove mathematics from the school curriculum?

SALVIATI: The mathematics has already been removed! The only question is what to do with the vapid, hollow shell that remains. Of course I would prefer to replace it with an active and joyful engagement with mathematical ideas.

SIMPLICIO: But how many math teachers know enough about their subject to teach it that way?

SALVIATI: Very few. And that's just the tip of the iceberg . . .

Mathematics in School

THERE IS SURELY NO MORE RELIABLE WAY TO KILL enthusiasm and interest in a subject than to make it a mandatory part of the school curriculum. Include it as a major component of standardized testing and you virtually guarantee that the education establishment will suck the life out of it. School boards do not understand what math is; neither do educators, textbook authors, publishing companies, and, sadly, neither do most of our math teachers. The scope of the problem is so enormous I hardly know where to begin.

Let's start with the "math reform" debacle. For many years there has been a growing awareness that something is rotten in the state of mathematics education. Studies have been commissioned, conferences assembled, and countless committees of teachers, textbook publishers, and educators (whatever they are) have been formed to "fix the problem." Quite apart from the self-serving interest paid to reform by the textbook industry (which profits from any minute political fluctuation by offering up "new" editions of their unreadable monstrosities), the entire reform movement has always missed the point. The mathematics curriculum doesn't need to be reformed, it needs to be *scrapped*.

All this fussing and primping about which "topics" should be taught in what order, or the use of this notation instead of that notation, or which make and model of calculator to use, for god's sake —it's like rearranging the deck chairs on the *Titanic*! Mathematics is *the music of reason*. To do mathematics is to engage in an act of discovery and conjecture, intuition and inspiration; to be in a state of confusion—not because it makes no sense to you, but because you *gave* it sense and you still don't understand what your creation is up to; to have a breakthrough idea; to be frustrated as an artist; to be awed and overwhelmed by an almost painful beauty; to be *alive*, damn it. Remove this from mathematics and you can have all the conferences you like; it won't matter. Operate all you want, doctors: *your patient is already dead*.

The saddest part of all this "reform" are the attempts to "make math interesting" and "relevant to kids' lives." You don't need to *make* math interesting—it's already more interesting than we can handle! And the glory of it is its complete *irrelevance* to our lives. That's why it's so fun!

Attempts to present mathematics as relevant to daily life inevitably appear forced and contrived: "You see, kids, if you know algebra then you can figure out how old Maria is if we know that she is two years older than twice her age seven years ago!" (As if anyone would ever have access to that ridiculous kind of information, and not her age.) Algebra is not about daily life, it's about numbers and symmetry—and this is a valid pursuit in and of itself:

> Suppose I am given the sum and difference of two numbers. How can I figure out what the numbers are themselves?

Here is a simple and elegant question, and it requires no effort to be made appealing. The ancient Babylonians enjoyed working on such problems, and so do our students. (And I hope you will enjoy thinking about it too!) We don't need to bend over backwards to give mathematics relevance. It has relevance in the same way that any art does: that of being a meaningful human experience.

In any case, do you really think kids even want something that is relevant to their daily lives? You think something practical like compound interest is going to get them excited? People enjoy fantasy, and that is just what mathematics can provide—a relief from daily life, an anodyne to the practical workaday world.

A similar problem occurs when teachers or textbooks succumb to cutesiness. This is where, in an attempt to combat so-called "math anxiety" (one of the panoply of diseases which are actually *caused* by school), math is made to seem "friendly." To help your students memorize formulas for the area and circumference of a circle, for example, you might invent a whole story about Mr. C, who drives around Mrs. A and tells her how nice his two pies are ($C = 2\pi r$) and how her pies are square ($A = \pi r^2$) or some such nonsense. But what about the *real* story? The one about mankind's struggle with the problem of measuring curves; about Eudoxus and Archimedes and the method of exhaustion; about the transcendence of pi? Which is more interesting— measuring the rough dimensions of a circular piece of graph paper, using a formula that someone handed you without explanation (and made you memorize and practice over and over), or hearing the story of one of the most beautiful, fascinating problems and one of the most brilliant and powerful ideas in human history? We're killing people's interest in *circles* for god's sake!

Why aren't we giving our students a chance to even hear about these things, let alone giving them an opportunity to actually do some mathematics, and to come up with their own ideas, opinions, and reactions? What other subject is routinely taught without any mention of its history, philosophy, thematic development, aesthetic criteria, and current status? What other subject shuns its primary sources—beautiful works of art by some of the most creative minds in history—in favor of third-rate text-book bastardizations?

The main problem with school mathematics is that there are no *problems*. Oh, I know what *passes* for problems in math classes, these insipid "exercises." "Here is a type of problem. Here is how to solve it. Yes it will be on the test. Do exercises 1-35 odd for homework." What a sad way to learn mathematics: to be a trained chimpanzee.

But a problem, a genuine honest-to-goodness natural human *question*—that's another thing. How long is the diagonal of a cube? Do prime numbers keep going on forever? Is infinity a number? How many ways can I symmetrically tile a surface? The history of mathematics is the history of mankind's engagement with questions like these, not the mindless regurgitation of formulas and algorithms (together with contrived exercises designed to make use of them).

A good problem is something you don't know how to solve. That's what makes it a good puzzle, and a good opportunity. A good problem does not just sit there in isolation, but serves as a springboard to other interesting questions. A triangle takes up half its box. What about a pyramid inside its three-dimensional box? Can we handle this problem in a similar way?

I can understand the idea of training students to master certain techniques—I do that too. But not as an end in itself. Technique in mathematics, as in any art, should be learned in context. The great problems, their history, the creative process—that is the proper setting. Give your students a good problem, let them struggle and get frustrated. See what they come up with. Wait until they are dying for an idea, *then* give them some technique. But not too much.

So put away your lesson plans and your overhead projectors, your full-color textbook abominations, your CD-ROMs and the whole rest of the traveling circus freak show of contemporary education, and simply do mathematics with your students! Art teachers don't waste their time with textbooks and rote training in specific techniques. They do what is natural to their subject—they get the kids painting. They go around from easel to easel, making suggestions and offering guidance:

> STUDENT: I was thinking about our triangle problem, and I noticed something. If the triangle is really slanted then it *doesn't* take up half its box! See, look:

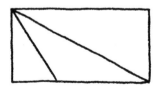

TEACHER: Excellent observation! Our chopping argument assumes that the tip of the triangle lies directly over the base. Now we need a new idea.

STUDENT: Should I try chopping it a different way?

TEACHER: Absolutely. Try all sorts of ideas. Let me know what you come up with!

So how do we teach our students to do mathematics? By choosing engaging and natural problems suitable to their tastes, personalities, and levels of experience. By giving them time to make discoveries and formulate conjectures. By helping them to refine their arguments and creating an atmosphere of healthy and vibrant mathematical criticism. By being flexible and open to sudden changes in direction to which their curiosity may lead. In short, by having an honest intellectual relationship with our students and our subject.

Of course what I'm suggesting is impossible for a number of reasons. Even putting aside the fact that statewide curricula and standardized tests virtually eliminate teacher autonomy, I doubt that most teachers even want to have such an intense relationship with their students. It requires too much vulnerability and too much responsibility—in short, it's too much work!

It is far easier to be a passive conduit of some publisher's "materials" and to follow the shampoo-bottle instruction—lecture, test, repeat—than to think deeply and thoughtfully about the meaning of one's subject and how best to convey that meaning directly and honestly to one's students. We are encouraged to forego the difficult task of making decisions based on our individual wisdom and conscience, and to get with the program. It is simply the path of least resistance:

TEXTBOOK PUBLISHERS : TEACHERS ::
(A) pharmaceutical companies : doctors
(B) record companies : disc jockeys
(C) corporations : congressmen
(D) all of the above

The trouble is that math, like painting or poetry, is *hard creative work*. That makes it very difficult to teach. Mathematics is a slow, contemplative process. It takes time to produce a work of art, and it takes a skilled teacher to recognize one. Of course it's easier to post a set of rules than to guide aspiring young artists, and it's easier to write a VCR manual than to write an actual book with a point of view.

Mathematics is an art, and art should be taught by working artists, or if not, at least by people who appreciate the art form and can recognize it when they see it. It is not necessary that you learn music from a professional composer, but would you want yourself or your child to be taught by someone who doesn't even play an instrument and has never listened to a piece of music in their lives? Would you accept as an art teacher someone who has never picked up a pencil or set foot in a museum? Why is it that we accept math teachers who have never produced an original piece of mathematics, know nothing of the history and philosophy of the subject, nothing about recent developments, nothing in fact beyond what they are expected to present to their unfortunate students? What kind of a teacher is that? How can someone teach something that they themselves don't do? I can't dance, and consequently I would never presume to think that I could teach a dance class (I could try, but it wouldn't be pretty). The difference is I *know* I can't dance. I don't have anyone telling me I'm good at dancing just because I know a bunch of dance words.

Now I'm not saying that math teachers need to be professional mathematicians—far from it. But shouldn't they at least understand what mathematics is, be good at it, and enjoy doing it?

If teaching is reduced to mere data transmission, if there is no sharing of excitement and wonder, if teachers themselves are passive recipients of information and not creators of new ideas, what hope is there for their students? If adding fractions is to the teacher an arbitrary set of rules, and not the outcome of a creative process and the result of aesthetic choices and desires, then *of course* it will feel that way to the poor students.

Teaching is not about information. It's about having an honest intellectual relationship with your students. It requires no method, no tools, and no training. Just the ability to be real. And if you can't be real, then you have no right to inflict yourself upon innocent children.

In particular, *you can't teach teaching*. Schools of education are a complete crock. Oh, you can take classes in early childhood development and whatnot, and you can be trained to use a blackboard "effectively" and to prepare an organized lesson plan (which, by the way, insures that your lesson will be *planned*, and therefore false), but you will never be a real teacher if you are unwilling to be a real person. Teaching means openness and honesty, an ability to share excitement, and a love of learning. Without these, all the education degrees in the world won't help you, and with them they are completely unnecessary.

It's perfectly simple. Students are not aliens. They respond to beauty and pattern, and are naturally curious like anyone else. Just talk to them. And more important, listen to them!

> SIMPLICIO: All right, I understand that there is an art to mathematics and that we are not doing a good job of exposing people to it. But isn't this a rather esoteric, highbrow sort of thing to expect from our school system? We're not trying to create philosophers here, we just want people to have a reasonable command of basic arithmetic so they can function in society.

> SALVIATI: But that's not true! School mathematics concerns itself with many things that have

nothing to do with the ability to get along in society—algebra and trigonometry, for instance. These studies are utterly irrelevant to daily life. I'm simply suggesting that if we are going to include such things as part of most students' basic education, that we do it in an organic and natural way. Also, as I said before, just because a subject happens to have some mundane practical use does not mean that we have to make that use the focus of our teaching and learning. It may be true that you have to be able to read in order to fill out forms at the DMV, but that's not why we teach children to read. We teach them to read for the higher purpose of allowing them access to beautiful and meaningful ideas. Not only would it be cruel to teach reading in such a way—to force third-graders to fill out purchase orders and tax forms—it wouldn't work! We learn things because they interest us now, not because they might be useful later. But this is exactly what we are asking children to do with math.

SIMPLICIO: But don't we need third-graders to be able to do arithmetic?

A Mathematician's Lament

SALVIATI: Why? You want to train them to calculate 427 plus 389? It's just not a question that very many eight-year-olds are asking. For that matter, most *adults* don't fully understand decimal place-value arithmetic, and you expect third-graders to have a clear conception? Or do you not care if they understand it? It is simply too early for that kind of technical training. Of course it can be done, but I think it ultimately does more harm than good. Much better to wait until their own natural curiosity about numbers kicks in.

SIMPLICIO: Then what *should* we do with young children in math class?

SALVIATI: Play games! Teach them chess and Go, Hex and backgammon, Sprouts and nim, whatever. Make up a game. Do puzzles. Expose them to situations where deductive reasoning is necessary. Don't worry about notation and technique; help them to become active and creative mathematical thinkers.

SIMPLICIO: It seems like we'd be taking an awful risk. What if we de-emphasize arithmetic so much that our students end up not being able to add and subtract?

SALVIATI: I think the far greater risk is that of creating schools devoid of creative expression of any kind, where the function of the students is to memorize dates, formulas, and vocabulary lists, and then regurgitate them on standardized tests—"Preparing tomorrow's workforce today!"

SIMPLICIO: But surely there is some body of mathematical facts of which an educated person should be cognizant.

SALVIATI: Yes, the most important of which is that mathematics is an art form done by human beings for pleasure! All right, yes, it would be nice if people knew a few basic things about numbers and shapes, for instance. But this will never come from rote memorization, drills, lectures, and exercises. You learn things by doing them and you remember what matters to you. We have millions of adults wandering around with "negative b plus or minus the square root of b squared minus $4ac$ all over $2a$" in their heads, and absolutely no idea whatsoever what it means. And the reason is that they were never given the chance to discover or invent such things for themselves. They never had an engaging problem to think about, to be frustrated by, and to create in them the desire for technique or method. They were never told the history of mankind's relationship with numbers—no ancient Babylonian problem tablets, no Rhind Papyrus, no *Liber Abaci*, no *Ars Magna*. More important, no chance for them to even get curious about a question; it was answered before they could ask it.

SIMPLICIO: But we don't have time for every student to invent mathematics for themselves! It took centuries for people to discover the Pythagorean theorem. How can you expect the average child to do it?

SALVIATI: I don't. Let's be clear about this. I'm complaining about the complete absence of art and invention, history and philosophy, context and perspective from the mathematics curriculum. That doesn't mean that notation, technique, and the development of a knowledge base have no place. Of course they do. We should have both. If I object to a pendulum being too far to one side, it doesn't mean I want it to be all the way on the other side. But the fact is, people learn better when the product comes out of the process. A real appreciation for poetry does not come from memorizing a bunch of poems, it comes from writing your own.

SIMPLICIO: Yes, but before you can write your own poems you need to learn the alphabet. The process has to begin somewhere. You have to walk before you can run.

SALVIATI: No, you have to have something you want to run *toward*. Children can write poems and stories *as* they learn to read and write. A piece of writing by a six-year-old is a wonderful thing, and the spelling and punctuation errors don't make it less so. Even very young children can invent songs, and they haven't a clue what key it is in or what type of meter they are using.

SIMPLICIO: But isn't math different? Isn't math a language of its own, with all sorts of symbols that have to be learned before you can use it?

SALVIATI: Not at all. Mathematics is not a language, it's an adventure. Do musicians speak another language simply because they choose to abbreviate their ideas with little black dots? If so, it's no obstacle to the toddler and her song. Yes, a certain amount of mathematical shorthand has evolved over the centuries, but it is in no way essential. Most mathematics is done with a friend over a cup of coffee, with a diagram scribbled on a napkin. Mathematics is and always has been about ideas, and a valuable idea transcends the symbols with which you choose to represent it. As Carl Friedrich Gauss once remarked, "What we need are *notions*, not notations."

SIMPLICIO: But isn't one of the purposes of mathematics education to help students think in a more precise and logical way, and to develop their quantitative reasoning skills? Don't all of these definitions and formulas sharpen the minds of our students?

SALVIATI: No, they don't. If anything, the current system has the effect of dulling the mind. Mental acuity of any kind comes from solving problems yourself, not from being told how to solve them.

SIMPLICIO: Fair enough. But what about those students who are interested in pursuing a career in science or engineering? Don't they need the training that the traditional curriculum provides? Isn't that why we teach mathematics in school?

SALVIATI: How many students taking literature classes will one day be writers? That is not why we teach literature, nor why students take it. We teach to enlighten everyone, not to train only the future professionals. In any case, the most valuable skill for a scientist or engineer is being able to think creatively and independently. The last thing anyone needs is to be *trained*.

The Mathematics Curriculum

T HE TRULY PAINFUL THING ABOUT THE WAY MATHE-matics is taught in school is not just what is missing—the fact that there is no actual math being done in our math classes—but what is there in its place: the confused heap of destructive disinformation known as "the mathematics curriculum." It is time now to take a closer look at exactly what our students are up against—what they are being exposed to in the name of mathematics, and how they are being harmed in the process.

The most striking thing about this so-called mathematics curriculum is its rigidity. This is especially true in the later grades. From school to school, city to city, and state to state, the exact same things are being said and done in the exact same way and in the exact same order. Far from being disturbed and upset by this Orwellian state of affairs, most people have simply accepted this standard model math curriculum as being synonymous with math itself.

This is intimately connected to what I call the "ladder myth"—the idea that mathematics can be arranged as a sequence of "subjects" each being in some way more advanced, or "higher," than the previous. The effect is to make school mathematics into a race—some students are "ahead" of others, and parents worry that their child is "falling behind." And where exactly does this race lead? What is waiting at the finish line? It's a sad race to nowhere. In the end you've been cheated out of a mathematical education, and you don't even know it.

Real mathematics doesn't come in a can—there is no such thing as an Algebra II *idea*. Problems lead you to where they take you. *Art is not a race.* The ladder myth is a false image of the subject, and a teacher's own path through the standard curriculum reinforces this myth and prevents him or her from seeing mathematics as an organic whole. As a result, we have a math curriculum with no historical perspective or thematic coherence, a fragmented collection of assorted topics and techniques, united only by the ease with which they can be reduced to step-by-step procedures.

In place of discovery and exploration, we have rules and regulations. We never hear a student saying, "I wanted to see if it could make any sense to raise a number to a negative power, and I found that you get a really neat pattern if you choose it to mean the reciprocal." Instead we have teachers and textbooks presenting the "negative exponent rule" as a fait accompli with no mention of the aesthetics behind this choice, or even that it is a choice.

In place of meaningful problems, which might lead to a synthesis of diverse ideas, to uncharted territories of discussion and debate, and to a feeling of thematic unity and harmony in mathematics, we have instead joyless and redundant exercises, specific to the technique under discussion, and so disconnected from each other and from mathematics as a whole that neither the students nor their teacher have the foggiest idea how or why such a thing might have come up in the first place.

In place of a natural problem context in which students can make decisions about what they want their words to mean, and what notions they wish to codify, they are instead subjected to an endless sequence of unmotivated and a priori definitions. The curriculum is obsessed with jargon and nomenclature, seemingly for no other purpose than to provide teachers with something to test the students on. No mathematician in the world would bother making these senseless distinctions: $2\frac{1}{2}$ is a "mixed number," while $\frac{5}{2}$ is an "improper fraction." They're *equal*, for crying out loud. They are the exact same numbers, and have the exact same properties. Who uses such words outside of fourth grade?

Of course it is far easier to test someone's knowledge of a pointless definition than to inspire them to create something beautiful and to find their own meaning. Even if we agree that a basic common vocabulary for mathematics is valuable, this isn't it. How sad that fifth-graders are taught to say "quadrilateral" instead of "four-sided shape," but are never given a reason to use words like "conjecture" and "counterexample." High school students must learn to use the secant function, 'sec x,' as an abbreviation for the reciprocal of the cosine function, '1 / cos x,' a definition with as much intellectual weight as the decision to use '&' in place of "and." That this particular shorthand, a holdover from fifteenth-century nautical tables, is still with us (whereas others, such as "versine," have died out) is mere historical accident, and is of utterly no value in an era when rapid and precise shipboard computation is no longer an issue. Thus we clutter our math classes with pointless nomenclature for its own sake.

In practice, the curriculum is not even so much a sequence of topics, or ideas, as it is a sequence of notations. Apparently mathematics consists of a secret list of mystical symbols and rules for their manipulation. Young children are given '+' and '÷.' Only later can they be entrusted with '$\sqrt{}$,' and then 'x' and 'y' and the alchemy of parentheses. Finally, they are indoctrinated in the use of 'sin,' 'log,' '$f(x)$,' and if they are deemed worthy, 'd' and '$\int$.' All without having had a single meaningful mathematical experience.

This program is so firmly fixed in place that teachers and textbook authors can reliably predict, years in advance, exactly what students will be doing, down to the very page of exercises. It is not at all uncommon to find second-year algebra students being asked to calculate $[f(x + h) - f(x)] / h$ for various functions f, so that they will have "seen" this when they take calculus a few years later. Naturally no motivation is given (nor expected) for why such a seemingly random combination of operations would be of interest, although I'm sure there are many teachers who try to explain what such a thing might mean, and think they are doing their students a favor, when in fact to them it is just one more boring math problem to be gotten over with. "What do they want me to do? Oh, just plug it in? OK."

Another example is the training of students to express information in an unnecessarily complicated form, merely because at some distant future period it will have meaning. Does any middle school algebra teacher have the slightest clue why he is asking his students to rephrase "the number x lies between three and seven" as $|x - 5| < 2$? Do these hopelessly inept textbook authors really believe they are helping students by preparing them for a possible day, years hence, when they might be operating within the context of a higher-dimensional geometry or an abstract metric space? I doubt it. I expect they are simply copying each other decade after decade, maybe changing the fonts or the highlight colors, and beaming with pride when a school system adopts their book and becomes their unwitting accomplice.

Mathematics is about problems, and problems must be made the focus of a student's mathematical life. Painful and creatively frustrating as it may be, students and their teachers should at all times be engaged in the process—having ideas, not having ideas, discovering patterns, making conjectures, constructing examples and counterexamples, devising arguments, and critiquing each other's work. Specific techniques and methods will arise naturally out of this process, as they did historically: not isolated from, but organically connected to, and an outgrowth of, their problem-background.

English teachers know that spelling and pronunciation are best learned in a context of reading and writing. History teachers know that names and dates are uninteresting when removed from the unfolding backstory of events. Why does mathematics education remain stuck in the nineteenth century? Compare your own experience of learning algebra with Bertrand Russell's recollection:

I was made to learn by heart: "The square of the sum of two numbers is equal to the sum of their squares increased by twice their product." I had not the vaguest idea what this meant and when I could not remember the words, my tutor threw the book at my head,

which did not stimulate my intellect in any
way.

Are things really any different today?

SIMPLICIO: I don't think that's very fair. Surely
teaching methods have improved since then.

SALVIATI: You mean *training* methods. Teaching
is a messy human relationship; it does not
require a method. Or rather I should say, if
you need a method you're probably not a
very good teacher. If you don't have enough
of a feeling for your subject to be able to talk
about it in your own voice, in a natural and
spontaneous way, how well could you understand it? And speaking of being stuck in the
nineteenth century, isn't it shocking how the
curriculum itself is stuck in the seventeenth?
To think of all the amazing discoveries and
profound revolutions in mathematical thought
that have occurred in the last three centuries!
There is no more mention of these than if they
had never happened.

SIMPLICIO: But aren't you asking an awful lot
from our math teachers? You expect them to

provide individual attention to dozens of students, guiding them on their own paths toward discovery and enlightenment, and to be up on recent mathematical history as well?

SALVIATI: Do you expect your art teacher to be able to give you individualized, knowledgeable advice about your painting? Do you expect her to know anything about the last three hundred years of art history? But seriously, I don't expect anything of the kind, I only wish it were so.

SIMPLICIO: So you blame the math teachers?

SALVIATI: No, I blame the culture that produces them. The poor devils are trying their best, and are only doing what they've been trained to do. I'm sure most of them love their students and hate what they are being forced to put them through. They know in their hearts that it is meaningless and degrading. They can sense that they have been made cogs in a great soul-crushing machine, but they lack the perspective needed to understand it, or to fight against it. They only know they have to get the students "ready for next year."

SIMPLICIO: Do you really think that most students are capable of operating on such a high level as to create their own mathematics?

SALVIATI: If we honestly believe that creative reasoning is too "high" for our students, and that they can't handle it, why do we allow them to write history papers or essays about Shakespeare? The problem is not that the students can't handle it, it's that none of the teachers can. They've never proved anything themselves, so how could they possibly advise a student? In any case, there would obviously be a range of student interest and ability, as there is in any subject, but at least students would like or dislike mathematics for what it really is, and not for this perverse mockery of it.

SIMPLICIO: But surely we want all of our students to learn a basic set of facts and skills. That's what a curriculum is for, and that's why it is so uniform—there are certain timeless, cold, hard facts we need our students to know: one plus one is two, and the angles of a triangle add up to 180 degrees. These are not opinions, or mushy artistic feelings.

SALVIATI: On the contrary. Mathematical structures, useful or not, are invented and developed within a problem context and derive their meaning from that context. Sometimes we want one plus one to equal zero (as in so-called 'mod 2' arithmetic) and on the surface of a sphere the angles of a triangle add up to more than 180 degrees. There are no facts per se; everything is relative and relational. It is the story that matters, not just the ending.

SIMPLICIO: I'm getting tired of all your mystical mumbo-jumbo! Basic arithmetic, all right? Do you or do you not agree that students should learn it?

SALVIATI: That depends on what you mean by "it." If you mean having an appreciation for the problems of counting and arranging, the advantages of grouping and naming, the distinction between a representation and the thing itself, and some idea of the historical development of number systems, then yes, I do think our students should be exposed to such things. If you mean the rote memorization of arithmetic facts without any underlying conceptual framework, then no. If you mean exploring the not-at-all obvious fact that five groups of seven is the same as seven groups of five, then yes. If you mean making a rule that $5 \times 7 = 7 \times 5$, then no. Doing mathematics should always mean discovering patterns and crafting beautiful and meaningful explanations.

SIMPLICIO: What about geometry? Don't students prove things there? Isn't high school geometry a perfect example of what you want math classes to be?

High School Geometry: Instrument of the Devil

THERE IS NOTHING QUITE SO VEXING TO THE AUTHOR of a scathing indictment as having the primary target of his venom offered up in his support. And never was a wolf in sheep's clothing as insidious, nor a false friend as treacherous, as high school geometry. It is precisely *because* it is school's attempt to introduce students to the art of argument that makes it so very dangerous.

Posing as the arena in which students will finally get to engage in true mathematical reasoning, this virus attacks mathematics at its heart, destroying the very essence of creative rational argument, poisoning the students' enjoyment of this fascinating and beautiful subject, and permanently disabling them from thinking about math in a natural and intuitive way.

The mechanism behind this is subtle and devious. The student-victim is first stunned and paralyzed by an onslaught of pointless definitions, propositions, and notations, and is then slowly and painstakingly weaned away from any natural curiosity or intuition about shapes and their patterns by a systematic indoctrination into the stilted language and artificial format of so-called "formal geometric proof."

All metaphor aside, geometry class is by far the most mentally and emotionally destructive component of the entire K–12 mathematics curriculum. Other math courses may hide the beautiful bird, or put it in a cage, but in geometry class it is openly and cruelly tortured. (Apparently I am incapable of putting all metaphor aside.)

What is happening is the systematic undermining of the student's intuition. A proof, that is, a mathematical argument, is a work of fiction, a poem. Its goal is to *satisfy*. A beautiful proof should explain, and it should explain clearly, deeply, and elegantly. A well-written, well-crafted argument should feel like a splash of cool water, and be a beacon of light—it should refresh the spirit and illuminate the mind. And it should be *charming*.

There is nothing charming about what passes for

proof in geometry class. Students are presented a rigid and dogmatic format in which their so-called "proofs" are to be conducted—a format as unnecessary and inappropriate as insisting that children who wish to plant a garden refer to their flowers by genus and species.

Let's look at some specific instances of this insanity. We'll begin with the example of two crossed lines:

Now the first thing that usually happens is the unnecessary muddying of the waters with excessive notation. Apparently, one cannot simply speak of two crossed lines; one must give elaborate names to them. And not simple names like 'line 1' and 'line 2,' or even '*a*' and '*b*.' We must (according to high school geometry) select random and irrelevant points on these lines, and then refer to the lines using the special "line notation."

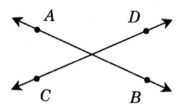

You see, now we get to call them $\overline{AB}$ and $\overline{CD}$. And god forbid you should omit the little bars on top—'AB' refers to the *length* of the line $\overline{AB}$ (at least I think that's how it works). Never mind how pointlessly complicated it is, this is the way one must learn to do it. Now comes the actual statement, usually referred to by some absurd name like:

PROPOSITION 2.1.1.

   Let $\overline{AB}$ and $\overline{CD}$ intersect at P.
   Then $\angle APC \cong \angle BPD$.

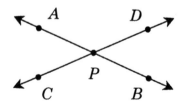

A Mathematician's Lament

In other words, the angles on both sides are the same. Well, duh! The configuration of two crossed lines is *symmetrical* for crissake. And as if this weren't bad enough, this patently obvious statement about lines and angles must then be "proved."

Proof:

STATEMENT	REASON
1. $m\angle APC + m\angle APD = 180$ $m\angle BPD + m\angle APD = 180$	1. Angle Addition Postulate
2. $m\angle APC + m\angle APD = m\angle BPD + m\angle APD$	2. Substitution Property
3. $m\angle APD = m\angle APD$	3. Reflexive Property of Equality
4. $m\angle APC = m\angle BPD$	4. Subtraction Property of Equality
5. $\angle APC \cong \angle BPD$	5. Angle Measurement Postulate

Instead of a witty and enjoyable argument written by an actual human being, and conducted in one of the world's many natural languages, we get this sullen, soulless, bureaucratic form-letter of a proof. And what a mountain being made of a molehill! Do we really want to suggest that a straightforward observation like this requires such an extensive preamble? Be honest: did you actually even read it? Of course not. Who would want to?

The effect of such a production being made over something so simple is to make people doubt their own intuition. Calling into question the obvious, by insisting that it be "rigorously proved" (as if the above even constitutes a legitimate formal proof), is to say to a student, "Your feelings and ideas are suspect. You need to think and speak our way."

Now there is a place for formal proof in mathematics, no question. But that place is not a student's first introduction to mathematical argument. At least let people get familiar with some mathematical objects, and learn what to expect from them, before you start formalizing everything. Rigorous formal proof only becomes important when there is a crisis—when you discover that your imaginary objects behave in a counterintuitive way; when there is a paradox of some kind. But such excessive preventative hygiene is completely unnecessary here— nobody's gotten sick yet! Of course if a logical crisis should arise at some point, then obviously it should be investigated, and the argument made more clear, but that process can be carried out intuitively and informally as well. In fact it is the soul of mathematics to carry out such a dialogue with one's own proof.

So not only are most kids utterly confused by this pedantry—nothing is more mystifying than a proof of the obvious—but even those few whose intuition remains intact must then retranslate their excellent, beautiful ideas back into this absurd hieroglyphic framework in order for their teacher to call it "correct." The teacher then flatters himself that he is somehow sharpening his students' minds.

As a more serious example, let's take the case of a triangle inside a semicircle:

Now the beautiful truth about this pattern is that no matter where on the circle you place the tip of the triangle, it always forms a nice right angle. (I have no objection to a term like "right angle" if it is relevant to the problem and makes it easier to discuss. It's not terminology itself that I object to, it's pointless, unnecessary terminology. In any case, I would be happy to use "corner" or even "pigpen" if a student preferred.)

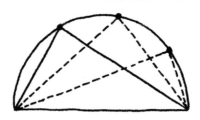

Here is a case where our intuition is somewhat in doubt. It's not at all clear that this should be true; it even seems *unlikely*—shouldn't the angle change if I move the tip? What we have here is a fantastic math problem! Is it true? If so, *why* is it true? What a great project! What a terrific opportunity to exercise one's ingenuity and imagination! Of course no such opportunity is given to the students, whose curiosity and interest is immediately deflated by:

THEOREM 9.5.
Let $\triangle ABC$ be inscribed in a semicircle
with diameter $\overline{AC}$.
Then $\angle ABC$ is a right angle.

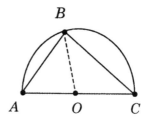

A Mathematician's Lament

Proof:

STATEMENT	REASON
1. Draw radius $\overline{OB}$. Then $OB = OC = OA$	1. Given
2. $m\angle OBC = m\angle BCA$ $m\angle OBA = m\angle BAC$	2. Isosceles Triangle Theorem
3. $m\angle ABC = m\angle OBA + m\angle OBC$	3. Angle Sum Postulate
4. $m\angle ABC + m\angle BCA + m\angle BAC = 180$	4. The sum of the angles of a triangle is 180
5. $m\angle ABC + m\angle OBC + m\angle OBA = 180$	5. Substitution (line 2)
6. $2\, m\angle ABC = 180$	6. Substitution (line 3)
7. $m\angle ABC = 90$	7. Division Property of Equality
8. $\angle ABC$ is a right angle	8. Definition of Right Angle

Could anything be more unattractive and inelegant? Could any argument be more obfuscatory and unreadable? This isn't mathematics! A proof should be an epiphany from the gods, not a coded message from the Pentagon. This is what comes from a misplaced sense of logical rigor: *ugliness*. The spirit of the argument has been buried under a heap of confusing formalism.

No mathematician works this way. No mathematician has *ever* worked this way. This is a complete and utter misunderstanding of the mathematical enterprise. Mathematics is not about erecting barriers between ourselves and our intuition, and making simple things complicated. Mathematics is about removing obstacles to our intuition, and keeping simple things simple.

Compare this unappetizing mess of a proof with the following argument devised by one of my seventh-graders:

> Take the triangle and rotate it around so it makes a four-sided box inside the circle. Since the triangle got turned completely around, the sides of the box must be parallel, so it makes a parallelogram. But it can't be a slanted box because both of its diagonals are diameters of the circle, so they're equal, which means it must be an actual rectangle. That's why the corner is always a right angle.

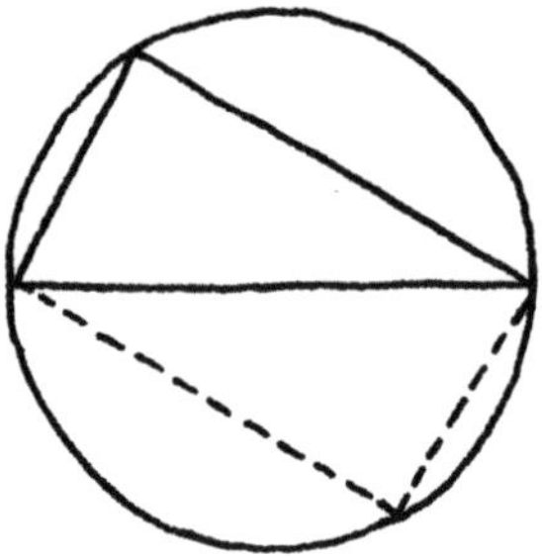

A Mathematician's Lament

Isn't that just delightful? And the point isn't whether this argument is any better than the other one *as an idea*, the point is that the idea comes across. (As a matter of fact, the idea of the first proof is quite pretty, albeit seen as through a glass, darkly.)

More important, the idea was the student's *own*. The class had a nice problem to work on, conjectures were made, proofs were attempted, and this is what one student came up with. Of course it took several days, and was the end result of a long sequence of failures.

To be fair, I did paraphrase the proof considerably. The original was quite a bit more convoluted, and contained a lot of unnecessary verbiage (as well as spelling and grammatical errors). But I think I got the feeling of it across. And these defects were all to the good; they gave me something to do as a teacher. I was able to point out several stylistic and logical problems, and the student was then able to improve the argument. For instance, I wasn't completely happy with the bit about both diagonals being diameters—I didn't think that was entirely obvious—but that only meant there was more to think about and more understanding to be gained from the situation. And in fact the student was able to fill in this gap quite nicely:

Since the triangle got rotated halfway around the circle, the tip must end up exactly opposite from where it started. That's why the diagonal of the box is a diameter.

So a great project and a beautiful piece of mathematics. I'm not sure who was more proud, the student or myself. This is exactly the kind of experience I want my students to have.

The problem with the standard geometry curriculum is that the private, personal experience of being a struggling artist has virtually been eliminated. The art of proof has been replaced by a rigid step-by-step pattern of uninspired formal deductions. The textbook presents a set of definitions, theorems, and proofs, the teacher copies them onto the blackboard, and the students copy them into their notebooks. They are then asked to mimic them in the exercises. Those that catch on to the pattern quickly are the "good" students.

The result is that the student becomes a passive participant in the creative act. Students are making statements to fit a preexisting proof-pattern, not because they *mean* them. They are being trained to ape arguments, not to *intend* them. So not only do they have no idea what their teacher is saying, *they have no idea what they themselves are saying*.

Even the traditional way in which definitions are presented is a lie. In an effort to create an illusion of clarity before embarking on the typical cascade of propositions and theorems, a set of definitions is provided so that statements and their proofs can be made as succinct as possible. On the surface this seems fairly innocuous; why not make some abbreviations so that things can be said more economically? The problem is that definitions matter. They come from aesthetic decisions about what distinctions you as an artist consider important. And they are *problem generated*. To make a definition is to highlight and call attention to a feature or structural property. Historically this comes out of working on a problem, not as a prelude to it.

The point is you don't start with definitions, you start with problems. Nobody ever had an idea of a number being "irrational" until Pythagoras attempted to measure the diagonal of a square and discovered that it could not be represented as a fraction. Definitions make sense when a point is reached in your argument which makes the distinction necessary. To make definitions without motivation is more likely to *cause* confusion.

This is yet another example of the way that students are shielded and excluded from the mathematical process. Students need to be able to make their own definitions as the need arises—to frame the debate themselves. I don't want students saying, "the definition, the theorem, the proof," I want them saying, "my definition, my theorem, my proof."

All of these complaints aside, the real problem with this kind of presentation is that it is boring. Efficiency and economy simply do not make good pedagogy. I have a hard time believing that Euclid would approve of this; I know Archimedes wouldn't.

> SIMPLICIO: Now hold on a minute. I don't know about you, but I actually *enjoyed* my high school geometry class. I liked the structure, and I enjoyed working within the rigid proof format.
>
> SALVIATI: I'm sure you did. You probably even got to work on some nice problems occasionally. Lots of people enjoy geometry class (although lots more hate it). But this is not a point in favor of the current regime. Rather, it

is powerful testimony to the allure of mathematics itself. It's hard to completely ruin something so beautiful; even this faint shadow of mathematics can still be engaging and satisfying. Many people enjoy paint-by-numbers as well; it is a relaxing and colorful manual activity. That doesn't make it the real thing, though.

SIMPLICIO: But I'm telling you, I *liked* it.

SALVIATI: And if you had had a more natural mathematical experience you would have liked it even more.

SIMPLICIO: So we're supposed to just set off on some free-form mathematical excursion, and the students will learn whatever they happen to learn?

SALVIATI: Precisely. Problems will lead to other problems, technique will be developed as it becomes necessary, and new topics will arise naturally. And if some issue never happens to come up in thirteen years of schooling, how interesting or important could it be?

SIMPLICIO: You've gone completely mad.

SALVIATI: Perhaps I have. But even working within the conventional framework, a good teacher can guide the discussion and the flow of problems so as to allow the students to discover and invent mathematics for themselves. The real problem is that the bureaucracy does not allow an individual teacher to do that. With a set curriculum to follow, a teacher cannot lead. There should be no standards, and no curriculum. Just individuals doing what they think best for their students.

SIMPLICIO: But then how can schools guarantee that their students will all have the same basic knowledge? How will we accurately measure their relative worth?

SALVIATI: They can't, and we won't. Just like in real life. Ultimately you have to face the fact that people are all different, and that's just fine. In any case, there's no urgency. So a person graduates from high school not knowing the half-angle formulas. (As if they do now!) So what? At least that person would come away with some sort of an idea of what the subject is really about, and would get to see something beautiful.

To put the finishing touches on my critique of the standard curriculum, and as a service to the community, I now present the first ever *completely honest* course catalog for K–12 mathematics:

THE STANDARD SCHOOL MATHEMATICS CURRICULUM

LOWER SCHOOL MATH. The indoctrination begins. Students learn that mathematics is not something you do, but something that is done to you. Emphasis is placed on sitting still, filling out worksheets, and following directions. Children are expected to master a complex set of algorithms for manipulating Hindu-Arabic symbols, unrelated to any real desire or curiosity on their part, and regarded only a few centuries ago as too difficult for the average adult. Multiplication tables are stressed, as are parents, teachers, and the kids themselves.

MIDDLE SCHOOL MATH. Students are taught to view mathematics as a set of procedures, akin to religious rites, which are eternal and set in stone. The holy tablets, or Math Books, are handed out, and the students learn to address the church elders as "they." (As in "What do they want here? Do they want me to divide?") Contrived and artificial "word problems" will be introduced in order to make the mindless drudgery of arithmetic seem enjoyable by comparison. Students will be tested on a wide array of unnecessary technical terms, such as 'whole number' and 'proper fraction,' without the slightest rationale for making such distinctions. Excellent preparation for Algebra I.

ALGEBRA I. So as not to waste valuable time thinking about numbers and their patterns, this course instead focuses on symbols and rules for their manipulation. The smooth narrative thread that leads from ancient Mesopotamian tablet problems to the high art of the Renaissance algebraists is discarded in favor of a disturbingly fractured, postmodern retelling with no characters, plot, or theme. The insistence that all numbers and expressions be put into various standard forms will provide additional confusion as to the meaning of identity and equality. Students must also memorize the quadratic formula for some reason.

GEOMETRY. Isolated from the rest of the curriculum, this course will raise the hopes of students who wish to engage in meaningful mathematical activity, and then dash them. Clumsy and distracting notation will be introduced, and no pains will be spared to make the simple seem complicated. The goal of this course is to eradicate any last remaining vestiges of natural mathematical intuition, in preparation for Algebra II.

ALGEBRA II. The subject of this course is the unmotivated and inappropriate use of coordinate geometry. Conic sections are introduced in a coordinate framework so as to avoid the aesthetic simplicity of cones and their sections. Students will learn to rewrite quadratic forms in a variety of standard formats for no reason whatsoever. Exponential and logarithmic functions are also introduced in Algebra II, despite not being algebraic objects, simply because they have to be stuck in somewhere, apparently. The name of the course is chosen to reinforce the ladder mythology. Why Geometry occurs in between Algebra I and its sequel remains a mystery.

TRIGONOMETRY. Two weeks of content are stretched to semester length by masturbatory definitional runarounds. Truly interesting and beautiful phenomena, such as the way the sides of a triangle depend on its angles, will be given the same emphasis as irrelevant abbreviations and obsolete notational conventions, in order to prevent students from forming any clear idea as to what the subject is about. Students will learn such mnemonic devices as "SohCahToa" and "All Students Take Calculus" in lieu of developing a natural intuitive feeling for orientation and symmetry. The measurement of triangles will be discussed without mention of the transcendental nature of the trigonometric functions, or the consequent linguistic and philosophical problems inherent in making such measurements. Calculator required, so as to further blur these issues.

PRE-CALCULUS. A senseless bouillabaisse of disconnected topics. Mostly a half-baked attempt to introduce late-nineteenth-century analytic methods into settings where they are neither necessary nor helpful. Technical definitions of limits and continuity are presented in order to obscure the intuitively clear notion of smooth change. As the name suggests, this course prepares the student for Calculus, where the final phase in the systematic obfuscation of any natural ideas related to shape and motion will be completed.

CALCULUS. This course will explore the mathematics of motion, and the best ways to bury it under a mountain of unnecessary formalism. Despite being an introduction to both the differential and integral calculus, the simple and profound ideas of Newton and Leibniz will be discarded in favor of the more sophisticated function-based approach developed as a response to various analytic crises that do not really apply in this setting, and that will of course not be mentioned. To be taken again in college, verbatim.

* * *

And there you have it. A complete prescription for permanently disabling young minds—a proven cure for curiosity. What have they done to mathematics!

There is such breathtaking depth and heartbreaking beauty in this ancient art form. How ironic that people dismiss mathematics as the antithesis of creativity. They are missing out on an art form older than any book, more profound than any poem, and

more abstract than any abstract. And it is *school* that
has done this! What a sad endless cycle of innocent
teachers inflicting damage upon innocent students.
We could all be having so much more fun.

SIMPLICIO: All right, I'm thoroughly depressed.
What now?

SALVIATI: Well, I think I have an idea about a
pyramid inside a cube . . .

PART II

Exultation

AND SO THE SENSELESS TRAGEDY KNOWN AS "mathematics education" continues, and only grows more indefensibly asinine and corrupt with each passing year. But I don't really want to talk about that anymore. I'm tired of complaining. And what's the point? School has never been about thinking and creating. School is about training children to *perform* so that they can be *sorted*. It's no shock to learn that math is ruined in school; *everything* is ruined in school! Besides, you don't need me to tell you that your math class was a boring, pointless waste of time—you went through it yourself, remember?

So what I'd rather do is tell you more about what math really is and why I love it so much. As I said before, the most important thing to understand is that mathematics is an art. Math is something you *do*. And what you are doing is exploring a very special and peculiar place—a place known as "Mathematical Reality." This is of course an imaginary place, a landscape of elegant, fanciful structures, inhabited by wonderful, imaginary creatures who engage in all sorts of fascinating and curious behaviors. I want to give you a feeling for what Mathematical Reality looks and feels like and why it is so attractive to me, but first let me just say that this place is so breathtakingly beautiful and entrancing that I actually spend a good part of my waking life there. I think about it all the time, as do most other mathematicians. We like it there, and we just can't stay away from the place.

In this way, being a mathematician is a lot like being a field biologist. Imagine that you have set up your camp on the outskirts of a tropical jungle, let's say in Costa Rica. Every morning you take your machete into the jungle and explore and make observations, and every day you fall more in love with the richness and splendor of the place. Suppose you are interested in a particular type of animal, say hamsters. (Let's not worry about whether there actually are any hamsters in Costa Rica.)

The thing about hamsters is they have *behavior*. They do cool, interesting things: they dig, they mate, they run around and make nests in hollow logs. Maybe you've studied a particular group of Costa Rican hamsters enough that you've tagged them and given them names. Maybe Rosie is black and white and loves to burrow; maybe Sam is brown and enjoys lying in the sun. The point is that you are watching, noticing, and getting *curious*.

Why do some hamsters behave differently from others? What features are common to all hamsters? Can hamsters be classified and grouped in meaningful and interesting ways? How do new hamsters get created from old ones, and what traits are inherited? In short, you've got hamster problems—natural, engaging questions about hamsters that you want answered.

Well, I've got problems too. Only they are not located in Costa Rica, and they don't concern hamsters. But the feeling is the same. There's a jungle full of strange creatures with interesting behaviors, and I want to understand them. For example, among my favorite denizens of the mathematical jungle are these fantastical beasts: 1, 2, 3, 4, 5, . . .

Please don't freak out on me here. I know you've probably had some pretty miserable experiences connected with these particular symbols, and I can feel your chest tightening already. Just relax. Everything is going to be fine. Trust me, I'm a doctor . . . of philosophy.

First of all, forget the symbols—they don't matter. Names never matter. Rosie and Sam do what they do; they don't care about your silly pet names for them. This is a hugely important idea: I'm talking about the difference between the thing itself and the *representation* of the thing. It is of absolutely no importance whatever what words you want to use (if any) or what symbols you wish to employ (if any). The only thing that matters in mathematics is what things *are*, and more important, how they *act*.

So somewhere along the line people started to *count* (no one knows quite when). A really big step occurred when people realized that they could represent things by other things (e.g., a caribou by a painting of a caribou, or a group of people by a pile of rocks). At some point (again, we don't know when) early humans conceived of the idea of *number*, of "three-ness" for instance. Not three berries, or three days, but three *in the abstract.* Throughout the millennia people have devised all sorts of languages for the representation of numbers—markers and tokens, coins with values on them, symbolic manipulation systems, and so on. Mathematically none of this really matters very much. From my point of view (that of the impractical daydreaming mathematician) a symbolic representation like '432' is no better or worse than an imaginary pile of four hundred thirty-two rocks (and in many ways I prefer the rocks). To me the important step is not the move from rocks to symbols, it's the transition from quantity to *entity*—the conception of five and seven not as amounts of something but as *beings*, like hamsters, which have features and behavior.

For example, to an algebraist such as myself, the statement 5 + 7 = 12 does not so much say that five lemons and seven lemons make twelve lemons (although it certainly does say that). What it says to me is that the entities commonly known by the nicknames "five" and "seven" like to engage in a certain activity (namely "adding") and when they do they form a new entity, the one we call "twelve." And this is what these creatures *do*—no matter what they are called or by whom. In particular, twelve does not "start with a one" or "end with a two." Twelve itself doesn't start or end, it just *is*. (What does a pile of rocks "start" with?) It is only the Hindu-Arabic decimal place-value *representation* of twelve that starts with a '1' and ends with a '2.' And that's really neither here nor there. Do you get what I'm saying?

As mathematicians we are interested in the *intrinsic* properties of mathematical objects, not the mundane features of some arbitrary cultural construct. The symbol '69' may look the same upside down, but the *number* sixty-nine doesn't "look" any way at all. I hope you can see how this point of view is a natural outgrowth of the "simple is beautiful" aesthetic. What do I care what notation system some Arabic traders introduced into Europe in the twelfth century? I care about my hamsters, not their names.

So let's try to think of these numbers 1, 2, 3, et cetera, as creatures with interesting behavior. Of course their behavior is determined by what they are, namely *sizes of collections*. (That's how we happened upon them in the first place!) Let's refer to them using imaginary piles of rocks:

This way we can observe them "in the wild," so to speak, and we won't be distracted or misled by some accidental artifact of notation. Now one behavior that people noticed pretty early on is that some of them (as piles of rocks) can be arranged in two equal rows:

The numbers four, eight, and fourteen have this property, whereas three, five, and eleven do not. And it's not because of their names—it's because of who they are and what they do. So here is a behavioral distinction among mathematical entities: some of them do this (the so-called "even" numbers) and some do not (the "odd" ones).

For pretty obvious reasons, I tend to think of even numbers as female and odd numbers as male. The even numbers (arranged in two equal rows) have a nice smooth profile, whereas the odd ones are always sticking something out:

Since pushing piles of rocks together is such a natural thing to do, it's also natural to wonder how the even/odd distinction is affected by addition. (It's like asking whether the spotted/plain trait in hamsters is inherited.) So I play around a bit with piles of rocks and I notice a lovely pattern:

> Even & Even makes Even
> Even & Odd makes Odd
> Odd & Odd makes Even

Do you see why? I especially like the way two odds fit together:

$$\overset{\text{OOO}}{\text{OO}} \;\&\; \overset{\text{OOO}}{\text{OOOO}} \;=\; \overset{\text{OOOOO}}{\text{OOOOOO}}$$

There's such a wonderful "two wrongs make a right" quality to this. Those annoying prongs just cancel each other out! And notice that this works for all odd numbers, not just the ones I happened to choose. In other words, this is a completely *general* behavior. So that's a nice discovery. Not that there's anything so special about using *two* rows. We could also investigate what happens when we arrange numbers into three rows, or four, or ten. What do our hamsters do then?

Now I know none of this is terribly sophisticated, but I really want you to get this feeling of imaginary entities and their amusing behavior. It's important for understanding both the attraction of the subject and its methodology (especially in the modern era). There is, however, an absolutely crucial difference between Costa Rican hamsters and mathematical entities like numbers or triangles: hamsters are *real*. They are part of physical reality.

Mathematical objects, even if initially inspired by some aspect of reality (e.g., piles of rocks, the disc of the moon), are still nothing more than figments of our imagination.

Not only that, but they are created by us and are endowed by us with certain characteristics; that is, they are what we ask them to be. Not that we don't build things in real life, but we are always constrained and hampered by the nature of reality itself. There are things I might want that I simply can't have because of the way atoms and gravity work. But in Mathematical Reality, because it is an imaginary place, I actually can have pretty much whatever I want. If you tell me, for instance, that $1 + 1 = 2$ and there's nothing I can do about it, I could simply dream up a new kind of hamster, one that when you add it to itself disappears: $1 + 1 = 0$. Maybe this '0' and '1' aren't collection sizes anymore, and maybe this "adding" isn't pushing collections together, but I still get a "number system" of a sort. Sure, there will be consequences (such as all even numbers being equal to zero), but so be it.

In particular, we are free to embellish or "improve" our imaginary structures if we see fit. For example, over the centuries it gradually dawned on mathematicians that this collection, 1, 2, 3, et cetera, is in some ways quite inadequate. There is actually a rather unpleasant asymmetry to this system, in that I can always add rocks but I can't always take them away. "You can't take three from two" is an obvious maxim of the real world, but we mathematicians do not like being told what we can and cannot do. So we throw in some new hamsters in order to make the system prettier. Specifically, after expanding our notion of collection sizes to include zero (the size of the *empty* collection), we can then define new numbers like '–3' to be "that which when added to three makes zero." And similarly for the other negative numbers. Notice the philosophy here—a number is what a number *does*.

In particular, we can replace the old-fashioned notion of subtraction by a more modern idea: *adding the opposite*. Instead of "eight take away five," we can (if we wish) view this activity as "eight plus negative five." The advantage here is that we have only one operation to deal with: adding. We have transferred the subtraction idea away from the world of operations and over to the numbers themselves. So instead of taking off my shoe, I can think of it as putting on my "anti-shoe." And of course my anti-anti-shoe would just be my shoe. Do you see the charm in this viewpoint?

Similarly, if multiplication is something you are interested in (that is, making repeated copies of piles of rocks), you might also notice an unpleasant lack of symmetry. What number triples to make six? Why, two of course. But what triples to make seven? There isn't any pile of rocks like that. How annoying!

Of course we're not really talking about piles of rocks (or anti-rocks). We're talking about an abstract imaginary structure *inspired by rocks*. So if we want there to be a number which when tripled makes seven, then we can simply build one. We don't even have to go out to the garage and get tools—we just "bring it into being" linguistically. We can even give it a name like '7/3' (a modified Egyptian shorthand for "that which when multiplied by three makes seven.") And so on. All of the usual "rules" of arithmetic are simply the consequences of these aesthetic choices. What are so often presented to students as a cold, sterile set of facts and formulas are actually the exciting and dynamic results of these new creatures interacting with each other—the patterns they play out as a result of their inborn linguistic "nature."

In this way we play and create and try to get closer to ideal beauty. A famous example from the early seventeenth century is the invention of projective geometry. Here the idea is to "improve" Euclidean geometry by removing *parallelism*. Putting aside the historical motivations behind this decision (which have to do with the mathematics of perspective), we can at least appreciate the fact that *in general* two straight lines intersect at a single point, and parallel lines break this pattern. To put it another way, two points always determine a line, but two lines don't always determine a point.

The bold idea was to add *new points* to the classical Euclidean plane. Specifically, we create one new point "at infinity" for each direction in the plane. All the parallel lines in that direction will now "meet" at this new point. We can imagine the new point to be infinitely far away in that direction. Of course, since every line goes off in two opposite directions, the new point must lie infinitely far away in *both* directions! In other words, our lines are now infinite loops. Is that a far out idea, or what?

Notice that we do get what we wanted: *every* pair of lines now meets at exactly one point. If they intersected before, then they still do; if they were parallel, they now intersect "at infinity." (To be complete, we should also add one more line, namely the one consisting of all the infinite points.) Now any two points determine a unique line, and any two lines determine a unique point. What a nice environment!

Does this sound to you like the ravings of a lunatic? I admit it takes some getting used to. Perhaps you object to these new points on the grounds that they're not really "there." But was the Euclidean plane there to begin with?

The point is that there is no reality to any of this, so there are no rules or restrictions other than the ones we care to impose. And the aesthetic here is very clear, both historically and philosophically: if a pattern is interesting and attractive, then it's good. (And if it means having to work hard to bend your mind around a new idea, so much the better.) Make up anything you want, so long as it isn't boring. Of course this is a matter of taste, and tastes change and evolve. Welcome to art history! Being a mathematician is not so much about being clever (although lord knows that helps); it's about being aesthetically sensitive and having refined and exquisite taste.

In particular, *contradiction* is usually regarded as rather boring. So at the very least we want our mathematical creations to be logically consistent. This is especially an issue when making extensions or improvements to existing structures. We are of course free to do as we wish, but usually we want to extend a system in such a way that the new patterns do not conflict with the old ones. (Such is the case with the arithmetic of negative numbers and fractions, for instance.) Occasionally, this compels us to make decisions we might otherwise not want to make, such as forbidding division by zero (if a number such as '1/0' were to exist, it would conflict with the nice pattern that multiplication by zero always makes zero). Anyway, as long as you are consistent, you can pretty much have whatever you want.

So the mathematical landscape is filled with these interesting and delightful structures that we have built (or accidentally discovered) for our own amusement. We observe them, notice interesting patterns, and try to craft elegant and compelling narratives to explain their behavior.

At least that's what I do. There certainly are people out there whose approach is quite different— practical-minded people who seek mathematical models of reality to help them make predictions or to improve some aspect of the human condition (or at least improve the balance sheet of their corporate sponsors). Well, I'm not one of those people. The only thing I am interested in using mathematics for is to have a good time and to help others do the same. And for the life of me I can't imagine a more worth-while goal. We are all born into this world, and at some point we will die and that will be that. In the meantime, let's enjoy our minds and the wonderful and ridiculous things we can do with them. I don't know about you, but I'm here to have *fun*.

Let's go a little deeper into the jungle, shall we? Now, you have to appreciate that people have been doing mathematics for quite some time (and rather intensely for the last three thousand years or so) and we have made a lot of amazing discoveries. Here is one I've always loved: What happens when you add up the first few odd numbers?

$$1 + 3 = 4$$
$$1 + 3 + 5 = 9$$
$$1 + 3 + 5 + 7 = 16$$
$$1 + 3 + 5 + 7 + 9 = 25$$

To the novice this may seem like a random jum-ble of numbers, but the sequence:

A Mathematician's Lament

$$4, 9, 16, 25, \ldots$$

is far from random. In fact, these are precisely the *square* numbers. That is, these are just the numbers of rocks you need to make a perfect square design:

So the square numbers stand out from the rest as having this particularly attractive property, which is why they get a special name. The list goes on indefinitely of course, since you could make a square design of any size. (These are imaginary rocks and we therefore have an inexhaustible supply.)

But this is remarkable! Why should adding up consecutive odd numbers always make a square? Let's investigate further:

$$1 + 3 + 5 + 7 + 9 + 11 + 13 = 49$$
(which is 7×7)

$$1 + 3 + 5 + 7 + 9 + 11 + 13 + 15 + 17 + 19 = 100$$
(which is 10×10)

It seems to keep happening! And it's utterly beyond our control. Either this is a true (and surprising and beautiful) feature of odd numbers or it isn't, and we simply have no say in the matter. We may have brought these creatures into existence (and that is a serious philosophical question in itself) but now they are running amok and doing things we never intended. This is the Frankenstein aspect of mathematics—we have the authority to *define* our creations, to instill in them whatever features or properties we choose, but we have no say in what behaviors may then ensue as a *consequence* of our choices.

Now I can't make you be curious about this discovery; you either are or you aren't. But at least I can tell you why I am. For one thing, adding up odd numbers seems like a very different sort of activity than making a square (i.e., multiplying a number by itself). These two ideas just don't seem to have much to do with each other. There's something a bit counterintuitive about this. I am drawn in by the possibility of a *connection*—a new, unforeseen relationship that will improve my intuition and perhaps permanently change the way I think about these objects. I suppose that's really a key part of it for me: I want to be *changed*. I want to be affected in a fundamental way.

A Mathematician's Lament

That's maybe the biggest reason why I do mathematics. Nothing I have ever seen or done comes close to having the transformative power of math. My mind gets blown pretty much every day.

Another thing to notice is that the collection of odd numbers is *infinite*. This always makes for awe and fascination. If in fact our pattern doesn't continue, how will we ever know? Checking the first million cases doesn't prove anything—it might conceivably fail for the very next number. And in fact there are thousands of simple questions about whole numbers that remain unsolved to this day—we simply don't know if the pattern continues or not.

So I wonder how you feel about this question of ours. Perhaps it's simply not your cup of tea. Still, I hope you can appreciate why I like it. Mostly I love the abstraction of it all, the sheer simplicity. This isn't some complicated congressional redistricting issue, or even a question about colliding electrons. It's about odd numbers, for god's sake. It's the ethereal purity, the "more universal than the universe" quality that is so attractive to me. These aren't hairy, smelly hamsters with bloodstreams and intestines; they're happy, free, lighter-than-air constructs of my imagination. And they are absolutely *terrifying*.

Do you get what I mean here? So simple they're scary? These aren't science-fiction aliens, these are *aliens*. And they're up to something, apparently. They seem to always add up to squares. But why? At this point what we have is a *conjecture* about odd numbers. We have discovered a pattern, and we think it continues. We could even verify that it works for the first trillion cases if we wanted. We could then say that it's true for all practical purposes, and be done with it. But that's not what mathematics is about. Math is not about a collection of "truths" (however useful or interesting they may be). Math is about reason and understanding. We want to know *why*. And *not* for any practical purpose.

Here's where the art has to happen. Observation and discovery are one thing, but *explanation* is quite another. What we need is a *proof*, a narrative of some kind that helps us to understand why this pattern is occurring. And the standards for proof in mathematics are pretty damn high. A mathematical proof should be an absolutely clear logical deduction, which, as I said before, needs not only to satisfy, but to satisfy beautifully. That is the goal of the mathematician: to explain in the simplest, most elegant and logically satisfying way possible. To make the mystery melt away and to reveal a simple, crystalline truth.

Now if you were my apprentice and we had more time together, I would send you off at this point to think and struggle and see what kind of explanation you could cobble together. (And of course if you want to stop reading right now and get to work on it, that would be fantastic.) Since my goal here is to give you a taste of mathematical beauty, I will instead simply show you a nice proof and see what you think of it.

So how does one go about proving something like this? It's not like being a lawyer, where the goal is to persuade other people; nor is it like a scientist testing a theory. This is a unique art form within the world of rational science. We are trying to craft a "poem of reason" that explains fully and clearly and satisfies the pickiest demands of logic, while at the same time giving us goosebumps.

Sometimes I like to imagine a Two-Headed Monster of mathematical criticism. The first head demands a logically airtight explanation, one with absolutely no gaps in the reasoning or any fuzzy "hand-waving." This head is a stickler, and is utterly merciless. We all hate its constant nagging, but in our hearts we know it is right. The second head wants to see simple beauty and elegance, to be charmed and delighted, to attain not just verification but a deeper level of understanding. Usually this is the more difficult head to satisfy. Anyone can be logical (and in fact, the validity of a deduction can even be checked mechanically) but to produce a real proof requires inspiration and epiphany of the highest order. Similarly, it's not that hard to draw an accurate portrait. One can develop an eye and master the technique. But to draw a portrait that *means* something, that conveys emotion and speaks to us—that's something else entirely. In short, our goal is to appease the Monster.

Not that it's so easy to get *any* proof off the ground. Most of us are so frustrated with our problems that we would gladly settle for the ugliest and clunkiest of arguments (assuming it is logically valid). At least we would then be sure that our conjecture is right and there won't be any counterexamples. But it is an unsatisfactory state of affairs, and it cannot last. As Hardy says, "there is no permanent place in the world for ugly mathematics." History shows that eventually (maybe centuries later) someone will surely uncover the *real* proof, the one that conveys not just a message, but a *revelation*.

A Mathematician's Lament

But how do we do it? Nobody really knows. You just try and fail and get frustrated and hope for inspiration. For me it's an adventure, a journey. I usually know more or less where I want to go, I just don't know how to get there. The only thing I *do* know is that I'm not going to get there without a lot of pain and frustration and crumpled-up paper.

So let's imagine that you've been playing with this problem for a while, and then at some point you have this realization: what the pattern is saying is that any square design can be broken into pieces which are just the odd numbers. So you try out some chopping ideas. Your first few attempts are successful, but have no real unity to them; they are random-seeming and do not generalize:

Then, all of a sudden, in one breathless heart-stopping moment, the clouds part and you can finally *see*:

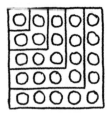

A square is a collection of nested L-shapes, and these L-shapes contain precisely the odd numbers. Eureka! Do you see why mathematicians jump out of bathtubs and run naked through the streets? Do you see why this useless, childish activity is so compelling?

The thing I want you especially to understand is this feeling of divine revelation. I feel that this structure was "out there" all along; I just couldn't see it. And now I can! This is really what keeps me in the math game—the chance that I might glimpse some kind of secret underlying truth, some sort of message from the gods.

To me, this kind of mathematical experience goes to the heart of what it means to be human. And I'll go even further and say that mathematics, this art of abstract pattern-making—even more than story-telling, painting, or music—is our most quintessentially human art form. This is what our brains do, whether we like it or not. We are biochemical pattern-recognition machines and mathematics is nothing less than the distilled essence of *who we are.*

Before we get too carried away, is it clear that these L-shapes do in fact follow the pattern? Is it so obvious that each successive L-shape contains exactly the next odd number, and that this pattern will continue forever? (This is the kind of skepticism typical of Head #1.) We know what we *think* these L-shapes are doing, and what we *want* them to do, but who says they will follow our desires?

This is something that happens in mathematics all the time. If proofs are stories, then they have parts, or episodes, like scenes in a novel. What our explanatory arguments do is break the problem down into subproblems. This is a big part of mathematical criticism. It's not that our proof is wrong or bad, we're just examining it more carefully, putting sections of it under the rational microscope.

So why do L-shapes make odd numbers? Of course the corner will always contain just one rock, and the next piece will have three, no matter how big the square is. Actually, I suppose we could entertain the possibility that our "square" consists of only one rock. It is up to you to decide if you want to include this sort of "trivial" case. The typical thing to do would be to include it, since it doesn't break the pattern: the sum of the first odd number, namely 1, is in fact the first square, 1×1. (If your taste goes further, and you want to include *zero*— being the sum of the first *none* odd numbers, and also 0×0—then you might want to seriously consider becoming a professional mathematician.) In any case, the first few L-shapes clearly comply with our wishes.

But is it clear that the pattern will keep going beyond our ability to draw pictures or to count? Let's imagine a hypothetical L-shape way down the line:

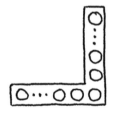

It is important to understand that I am not committing myself to any particular size here, but keeping my mind open and arguing *generally*—this is any size L-shape; the *n*th one if you will; the *generic*

one. Hopefully, we would then experience our next moment of clarity:

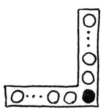

Any L-shape can be broken up into two "arms" and a "joint." The two arms are equal, so they contain the same number, and the joint adds one more. That's why the total is always odd! And what's more, when we go from one L-shape to the next, we see that each arm gets larger by exactly one:

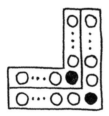

This means that each successive L-shape is exactly two more than the previous. And that's why the pattern keeps going!

So there's an example of what it's like to do mathematics. Playing with patterns, noticing things, making conjectures, searching for examples and counterexamples, being inspired to invent and explore, crafting arguments and analyzing them, and raising new questions. That's what it's all about. I'm not saying it's vitally important; it isn't. I'm not saying it will cure cancer; it won't. I'm saying it's fun and it makes me feel good. Plus, it's perfectly harmless. And how many human activities can you say *that* about?

Let me make a couple of important points. First of all, notice that once we know *why* something is true, then in particular we know *that* it is true. A trillion instances tells us nothing; when it comes to infinity, the only way to know what is to know why. Proof is our way of capturing an infinite amount of information in a finite way. That's really what it means for something to have a pattern—if we can capture it with *language*.

Another thing I want you to appreciate is the *finality* of mathematical proof. There's nothing tentative or hypothetical here. It's not going to turn out later that we were wrong. The argument is completely self-contained; we're not awaiting any experimental confirmation.

A Mathematician's Lament

Finally, I want to stress again that it's not the *fact* that consecutive odd numbers add up to squares that really matters here; it's the discovery, the explanation, the analysis. Mathematical truths are merely the incidental by-products of these activities. Painting is not about what hangs in the museum, it's about what you *do*—the experience you have with brushes and paint.

As I see it, art is not a collection of nouns, it's a *verb*—a way of life, even (or at any rate a means of escape). To reduce the adventure that we just went through together to a mere statement of fact would be to miss the point entirely. The point was that we *made* something. We made something beautiful and compelling and we had fun doing it. For a brief shining moment we lifted the veil and glimpsed a timeless simple beauty. Is this not something of value? Is humankind's most fascinating and imaginative art form not something worth exposing our children to? I think it is.

So let's do some math right now! We just saw that adding consecutive odd numbers always makes a square (and more important, we figured out why). What happens if we add up consecutive *even* numbers? How about adding up all the numbers? Is there a simple pattern? Can you explain why it happened? Have fun!

Now hold on a minute, Paul. Are you telling me that mathematics is nothing more than an exercise in mental masturbation? Making up imaginary patterns and structures for the hell of it and then investigating them and trying to devise pretty explanations for their behavior, all for the sake of some sort of rarified intellectual aesthetic?

Yep. That's what I'm saying. In particular, pure mathematics (by which I mean the fine art of mathematical proof) has absolutely no practical or economic value whatsoever. You see, practical things don't require explanation. Either they work or they don't. Even if you could find a way to put our odd number discovery to some sort of practical use (and of course there's lots of math out there that is indeed extremely useful) you would have no need for our gorgeous explanation. If it works for the first trillion numbers, then it works. Issues involving infinity simply don't come up in business or medicine.

Anyway, the point is not whether mathematics has any practical value—I don't care if it does or not. All I'm saying is that we don't need to *justify* it on that basis. We're talking about a perfectly innocent and delightful activity of the human mind—a dialogue with one's own mentality. Math requires no pathetic industrial or technological excuses. It transcends all of those mundane considerations. The value of mathematics is that it is fun and amazing and brings us great joy. To say that math is important because it is useful is like saying that children are important because we can train them to do spiritually meaningless labor in order to increase corporate profits. Or is that in fact what we *are* saying?

Let's quickly escape back to the jungle. Now just as hamsters occupy a certain biological niche—plants and insects they like to eat, geographic areas and terrain they inhabit—math problems are also situated within an environment—a *structural* environment. Let me try to illustrate this idea with another personal favorite.

Here are two points on one side of a straight line. The question is, what is the shortest path from one point to the other that touches the line? (Naturally, the part about touching the line is the interesting part—if we dropped that requirement then the answer would obviously be just the straight line connecting the two points.)

Clearly the shortest path must look something like this:

Since our path has to hit somewhere, we can't do better than to go straight there. The question is, where is "there"? Among all the possible points on the line, which one gives us the shortest path? Or could it be that they all have the same length?

What an elegant and fascinating problem! What a delightful setting in which to exercise our creativity and ingenuity. And notice: we don't even have a conjecture. We have no clue what the shortest path is, so we don't even know what we are trying to prove! So here we will have to discover not only an explanation for the truth, but what the truth is in the first place.

Again, the right thing for me to do as your math teacher would be *nothing*. That's a thing most teachers (and adults generally) seem to have a hard time doing. Were you my student (and assuming this problem interested you) I would simply say, "Have fun. Keep me posted." And your relationship to the problem would develop in whatever way it would.

Instead, I will use this opportunity to show you another lovely mathematical argument, which I hope will both charm and inspire you.

So it turns out that there is in fact only one shortest path and I will tell you how to find it. For convenience, let's give the points names, say *A* and *B*. Suppose we had a path from *A* to *B* that touches the line:

There's a very simple way to tell if such a path is as short as possible. The idea, which is one of the most surprising and unexpected in all of geometry, is to look at the *reflection* of the path across the line! To be specific, let's take one part of the path, say from where it hits the line to where it hits the point *B*, and reflect that part over the line:

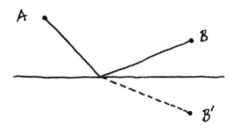

We now have a new path that starts at *A*, crosses the line, and ends up at the point *B'*, the reflection of the original point *B*. In this way, *any* path from *A* to *B* can be transformed into a path from *A* to *B'*:

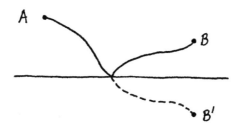

Now here's the point: the new path has exactly the same length as the original. Do you see why? This means that the problem of finding the shortest path from *A* to *B* that hits the line is the same as finding the shortest path from *A* to *B'*. But that's easy—it's just a straight line! In other words, the path we're looking for is simply the path that *when reflected* becomes straight!

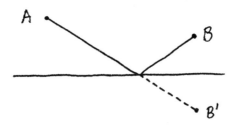

Is that great, or what? I only wish I could see your face—to see if your eyes light up, and to make sure that you get the joke, so to speak. Mathematics is fundamentally an act of communication, and I want to know if my idea got through. (If tears aren't streaming down your face, maybe you should read it again.)

I want you to know that when I first saw this proof I was absolutely shocked. The thing that got to me (and still does) is the *perversity* of it. The points were both above the line. Their shortest path is also. What the hell does this have to do with anything below the line? It was a shattering argument for me; definitely one of my formative mathematical experiences.

So I want to use this problem to make a few comments about the way modern mathematicians view their subject. What is this problem really about? What are the issues here? Well the first thing to notice is the *setting*—points, lines, a plane on which the action takes place, a sense of distance or length— these are the hallmarks of *geometric* structure. This problem fits into a larger category of problems concerned with spatial environments and notions of distance. These can range from the "elementary" geometric ideas of the classical Greeks (which were themselves inspired by earlier Egyptian practical observations about the real world) to the most abstract and bizarre imaginary structures—many having nothing whatever to do with anything even vaguely resembling reality. (Not that we know what reality is, but you get what I mean.)

Essentially, the adjective "geometric" is used by mathematicians to group together those problems and theories that concern some sort of collection of "points" (which may be quite arbitrary and abstract) and some sort of notion of "distance" between them (which also may bear no resemblance to anything familiar). For example, the "space" consisting of all red and blue bead strings of length five can be given a geometric structure by defining the distance between two such strings to be the number of places in the bead sequence where the colors disagree. Thus, the distance between the points 'RBBRB' and 'BBBRR' would be 2, since they differ only in the first and last places. Can you find an "equilateral triangle" (i.e., three points that all have the same distances to each other) inside of this space?

Similarly, problems can be classified as having algebraic, topological, or analytic structure, as well as many other types, and of course combinations of the above. Some areas of mathematics, such as the theory of sets or the study of order types, concern objects with almost no structure at all, whereas others (e.g., elliptic curves) involve practically every structural category under the sun. The point of this sort of framework is the same as it is in biology: to help us understand. Knowing that hamsters are mammals (and this is not an arbitrary classification, but a structural one) helps us make predictions and to know what to look out for. Classifications are a guide for our intuition. Similarly, knowing that our problem has geometric structure may give us fruitful ideas and keep us from wasting our time on approaches that are not in harmony with that structural world.

For example, any plan of attack on our shortest-path problem that involves bending or twisting is almost automatically doomed to fail, since such activities tend to distort shapes and mess up length information. We should instead think about activities and transformations that are *structure preserving*. In the case of our problem, which takes place in a Euclidean geometric environment, the natural activities would be those that preserve distances—namely sliding, rotating, and reflecting. From this perspective, the use of reflection maybe doesn't seem quite so shocking anymore; it is a natural element of the structural framework of the problem.

But that's not all. The thing about proofs is they always manage to prove more than you intended. The essence of the argument is the fact that reflection across a line preserves distances. This means that our argument applies to *any* setting in which there is a notion of point, line, distance and reflection. For instance, on the surface of a sphere there is a notion of reflection across an *equator*:

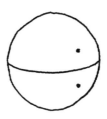

This means that equators (the curves you get when you chop a sphere in half) are the natural spherical analogs of "straight line." And in fact it happens to be true that the shortest path between two points on the surface of a sphere is to follow an equator (which is why airplanes often take such routes).

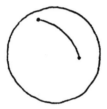

So the corresponding problem on a sphere would be: given two points on the same side of an equator, what is the shortest path between them that touches the equator? My point is that our exact same argument still works. Again it is the path that when reflected is straight:

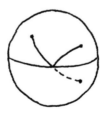

How about if we have two points in space on the same side of a plane?

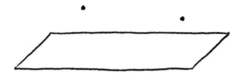

What I'm saying is that proofs are bigger than the problems they come from. A proof tells you what really matters and what is mere fluff, or irrelevant detail; it separates the wheat from the chaff. Of course, some proofs are better than others in this regard. Often a new argument is discovered that shows that what was previously thought to be an important assumption is in fact unnecessary. I suppose what I'm really trying to say here is that mathematical structures are designed and built not so much by us, as by *our proofs*.

The historical development of mathematics (especially in the past couple of centuries) exhibits a consistent, undeniable pattern: first come the problems, whose sources are many and varied, often inspired by the real world. Eventually, *connections* are made between diverse problems, usually due to common elements that appear in various proofs. Abstract structures are then devised that can "carry" the kind of information that forms the connection (the classic example being the "group" concept, which captures abstractly the idea of a closed system of activities, e.g., algebraic operations like addition, or systems of geometric or combinatorial transformations such as rotation or permutation). New questions then arise concerning the behavior of the new abstract structures—classification problems, construction of invariants, structure of sub-objects, et cetera. And the process continues with the discovery of new connections among the abstract structures themselves, generating even more powerful abstractions. Thus mathematics moves further and further away from its "naïve" origins. Some areas of mathematics, such as logic and category theory, concern themselves with spaces (so to speak) whose "points" are themselves mathematical theories!

As a small example, the key idea in our path problem was reflection. Now reflections have the amusing property that when you do them twice it's as if you've done nothing at all. Does that remind you of anything? It's just like our self-annihilating hamster—that new version of 1 with $1 + 1 = 0$. So here we have a connection between an algebraic structure and a geometric one. This raises a lot of questions concerning the extent to which number systems of various kinds can possess geometric "representations." Can you make up a number system that behaves like the rotations of a triangle?

All I'm really trying to say here is that as modern mathematicians we are always on the lookout for structure and structure-preserving transformations. This approach not only gives us a meaningful way to group problems together and to understand what they are really "about," but it also helps us to narrow the search for proof ideas. If a new problem comes along that lies in the same structural category as one we have already solved, we may be able to use or modify our previous methods.

Ok, grab your machete. It's back to the jungle we go! I can't resist giving you at least one more example of the mathematical aesthetic. This is what I like to call the "Friends at a Party" problem: Must there always be two people at a party who have the same number of friends there?

The first thing is to decide what we want our words to mean. What are people? What is friendship?

What exactly is a party? How does a mathematician address these issues? Surely we don't want to deal with actual humans and their complicated social lives. The aesthetic of simplicity demands that we shed all such unnecessary complexity and get to the heart of the matter. This is not a question about people and friendship, it's a question about relationships *in the abstract*. A party then becomes a "relationship structure" consisting of a set of objects (it doesn't matter what they are) together with a collection of (presumably mutual) relationships between them.

If we wanted, we could visualize such a structure using a simple diagram:

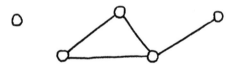

Here is a party of five, including one stranger (no friends) and a rather popular fellow with three friends. And it just so happens that there *are* two objects with the same number of connections (namely two).

A Mathematician's Lament

So here is a simple and beautiful class of mathematical structures (known in the math biz as combinatorial graphs) and a natural and amusing question about them: Does *every* graph possess a pair of objects with the same number of connections? (We're assuming of course that our graphs involve more than one object.)

So where do math problems like these come from? Well, I'll tell you: they come from *playing*. Just playing around in Mathematical Reality, often with no particular goal in mind. It's not hard to find good problems—just go to the jungle *yourself*. You can't take three steps without tripping over something interesting:

YOU: So Paul, I was thinking about what you said before about arranging numbers in rows, and I noticed that some numbers are so awkward they can't be arranged evenly in *any* number of rows. Like thirteen—it just doesn't work.

ME: Well, you could always arrange it as one row of thirteen . . . or as thirteen rows of one!

YOU: Yes, but that's *boring*. You can do that with any number. I'm talking about using at least two rows. So anyway, I started making a list of these weird numbers. It goes like this:

1, 2, 3, 5, 7, 11, 13, 17, 19, 23, 29, 31, 37, 41, 43, 47, . . .

and it seems to keep on going, but I haven't found any real pattern to it.

ME: Well, you've stumbled onto something very mysterious. The truth is, we don't know very much about these weird numbers of yours. One thing we *do* know is that they go on forever—there is an infinite supply of numbers that can't be arranged in rows. Maybe that would be a good thing for you to try to prove.

YOU: Yes, I'd like to think about it. Anyway, the thing I noticed about my list is the spacing between the numbers. It seems like they mostly thin out as they get bigger, but then sometimes you get these little clumps like 17, 19 and 101, 103 where they only jump by two. Does that keep happening?

ME: Nobody knows! Your weird numbers are called "primes" and the ones that come in pairs are called "prime twins." Your question about whether they keep occurring is known as the twin prime conjecture. It is actually one of the most famous unsolved problems in arithmetic. Most people who have worked on it (including myself) feel that it is probably true—prime twins should keep happening— but nobody knows for sure. I'm hoping to see a proof before I die, but I'm not terribly optimistic.

YOU: How bizarre that something so simple should turn out to be so hard! The other thing I noticed is that after 3, 5, 7 you never seem to get three primes in a row. Is that true?

ME: Prime triplets! What a terrific problem for you. Why don't you work on that and we'll see what you come up with . . .

(*A few days later*)

YOU: I think I've discovered something! I was looking for prime triplets, and what I noticed is that whenever you have three odd numbers in a row, one of them is *always* a multiple of three. Like with 13, 15, 17, the middle number is 5×3.

ME: That's fantastic! And it certainly explains why 3, 5, 7 is the last of the prime triplets— the only prime which is a multiple of three is three itself. So now you just have to figure out *why* three odds in a row must always contain a multiple of three.

YOU: Does this process ever stop? Does math ever come to an end?

ME: No, because solving problems always leads to new problems. For instance, now you've got me wondering whether five odd numbers in a row must always contain a multiple of five . . .

This is how math problems arise—just from sincere and serendipitous exploration. And isn't that how *every* great thing in life works? Children understand this. They know that learning and playing are the same thing. How sad that the grownups have forgotten. They think of learning as a chore, so they make it into one. Their problem is *intentionality*.

A Mathematician's Lament

So let me leave you with the only practical advice I have to offer: just play! You don't need a license to do math. You don't need to take a class or read a book. Mathematical Reality is *yours* to enjoy for the rest of your life. It exists in your imagination and you can do whatever you want with it. Including nothing, of course.

If you happen to be a student in school (and you have my condolences), then try to ignore the pointless absurdity of your math class. If you want, you can escape from the tedium by *actually doing mathematics*. It's nice to have interesting things to think about while you're staring out the window and waiting for the bell to ring.

And if you are a math teacher, then you *especially* need to be playing around in Mathematical Reality. Your teaching should flow naturally from your *own* experience in the jungle, not from some fake tourist version with a car on tracks and the windows rolled up. So throw the stupid curriculum and textbooks out the window! Then you and your students can start doing some *math* together. And seriously, if you have no interest in exploring your own personal imaginary universe, in making discoveries and trying to understand them, then what are you doing calling yourself a math teacher? If you don't have a personal relationship to your subject, and if it doesn't move you and send chills down your spine, then you need to find something else to do. If you love working with children and you really want to be a teacher, that's wonderful—but teach something that actually means something to you, about which you have something to say. It's important that we be honest about that. Otherwise I think we teachers can do a lot of unintentional harm.

And if you are neither student nor teacher, but simply a person living in this world and searching as we all are for love and meaning, I hope I have managed to give you a glimpse of something beautiful and pure, a harmless and joyful activity that has brought untold delight to many people for thousands of years.

Bellevue Literary Press is devoted to publishing literary fiction and nonfiction at the intersection of the arts and sciences because we believe that science and the humanities are natural companions for understanding the human experience. We feature exceptional literature that explores the nature of perception and the underpinnings of the social contract. With each book we publish, our goal is to foster a rich, interdisciplinary dialogue that will forge new tools for thinking and engaging with the world.

To support our press and its mission, and for our full catalogue of published titles, please visit us at blpress.org.

BELLEVUE LITERARY PRESS
New York